The Secret Life

of

Bar Codes

by John Berry

ISBN   978-0-9576032-0-2

Published by Wirksworth Books
Gatehouse Drive, Wirksworth
DE4 4DL, England

Printed by createspace

Acknowledgements

Many individuals from the world of bar coding in GS1, EAN and UCC, both past and present, have assisted me with their encouragement, detailed knowledge, suggestions and friendship. The book could not have been written without them. Hundreds of them have shared their passion, views and time with me during the past thirty years. A smaller but large number of them have been directly involved with this book. To highlight any names seems unfair to others not mentioned. So can I first offer my grateful thanks to all of you.

However, I must mention the staff of GS1 in Brussels who have offered patient support and prompted me to complete the book to coincide with the 40th anniversary of the selection of the bar code you see on products today. And I must single out Etienne Boonet and Andrew Osborne for the advice and encouragement they have given over many years and for the many factual corrections they made to my draft offerings. Any remaining errors which even their eagle eyes have not detected are my responsibility.

Then I must thank those who, with little previous knowledge of the subject, provided me with a wholly different set of comments. They corrected my typing, punctuation and use of language. But much more than that they told me what were the most boring and incomprehensible parts of the book. Again I would highlight two individuals who produced the most changes to my work, Jeremy Taylor and, with markings on every page, my daughter Christine. As you read the remaining pages of boredom and confusion, you can share my gratitude to them for what might have been.

After all those efforts, the style, format, detailed selection of what should be included and excluded, and views on the value of bar codes and predictions for the future are entirely my own.

Finally I am eternally grateful to my wife June. She sacrificed much over the 20 years and many days spent away from home in locations all over the world committed to this subject. She continued to encourage me when I retired and decided to write this book on our return from my last meeting in Beijing in 2000. It is my lasting regret that the combination of my own tardiness in completing it and the advancement of her Alzheimers Disease mean that she cannot fully share the pleasure of its final publication.

John Berry, Wirksworth 2013

Table of Contents

INTRODUCTION

PART I - The Old Retail World

PART II – The New Retail World (The Tip of the Iceberg)

PART III – The New Retail World (The Iceberg)

PART IV – Beyond the Retail World

APPENDICES

- EAN/GS1 Membership
- GS1 Officers 1977 – 2013
- Where did ENA/GS1 meet
- Significant milestones of GS1 History
- EAN/GS1 first scanning stores
- Alphabetical list of EAN.UCC country prefix allocations

Introduction

The manager worked for IBM. And he knew he was right.

Paul, this will never work. You're going to print these lines as a bar code on a label. You're going to attach this bar code label to this packet of cigarettes. You're going to pass this bar code label symbol by this open hole at a variable speed by a human hand. You're going to read the code. And you're going to do the same thing on 15 different checkouts in a store at the same time.

Then you're going to look a price up automatically using a computer in the back of the store and come back and print it out. And you're going to do this before the customer has got fed up with waiting and left the store! And you want me to pay for you to develop equipment that will do this?

The problem was that the person he was speaking to also worked for IBM. And he knew he was right too. And he was very very persuasive. So ...

OK. OK. I'll give you a year. I'll fund you for that time. But if this thing doesn't work, you can move your desk into the parking lot.

He came back – a year to the day. He looked at the scanner. He picked up the packet of cigarettes, with its bar code label, and tossed it from one end of the checkout to the other. As the packet bounced off the end, the printer broke the silence and .... printed the price of the packet of cigarettes.

He did not believe it. He knew the tricks that these technical people could get up to. He looked underneath the desk that the equipment was sitting on, because he wanted to be sure that they didn't have one of their engineers hiding there and keying the right answer in! But it really was true. It really did work.

In 1969, Bob Evans and Paul McEnroe, working for IBM at Raleigh in North Carolina could not have envisaged the sheer scale of the change to retailing throughout the world that was about to occur. They were part of a retail revolution. Look at what has changed in the last 44 years.

In 1969 shopping was so simple. The price of every item you bought was on the product. Sometimes it was printed on the packaging, but most of the time it was found on a small price sticker printed and attached to each item in the store. At the service counter or checkout, the cashier read each price sticker, keyed in each price on a cash register, and produced a till receipt with a list of these prices. Finally you paid for the goods in cash, with 'real money'. It was simple – but not without its problems.

The accuracy and efficiency of the system depended upon every price being found correctly in a price list and printed correctly on a price sticker by back-store staff, every single product taken to the checkout actually having a price sticker (or pre-printed price), and every price being key-entered correctly by the till operator. It would be true to say that this did not always happen. There was, to put it kindly, some scope for human error.

But in 2013 it is so different. The price of almost every item you buy is not on the product, but rather on the shelf-edge where the product is displayed. At the checkout, the operator takes each price-less item and 'reads' the bar code on each product with some form of scanner, which allows the price to be retrieved almost instantaneously from a computer in the store. The till receipt prints not only a list of prices but also a description of each item. In addition it details any discounts from special offers such as '3 for the price of 2', 'buy X and get Y at half price', '100 extra storecard points' etc. Finally, you probably pay for the goods with 'plastic money' and the amount is authorised in a matter of seconds by your bank's computer which could be located anywhere in the world.

In 1969, a bar code label was printed and stuck on the packet of cigarettes. In 2013, advances in printing technology mean that bar codes can be printed directly onto the packaging surface of any product – paper, card, plastic, glass or metal.

In 1969, it was hoped that it would be possible for 15 scanners in a store to read these bar codes simultaneously. Today we have large department stores with over 200 scanners reading bar codes and supermarkets handling over 3000 products per minute.

In this time, the use of bar codes and the item numbers printed below them has expanded into over 100 countries. There are no accurate figures for the sheer scale of their use. But a conservative estimate would say that that they are on over 50 million different products sold by over 1 million companies which are read in over 500,000 retail stores. They are used far beyond the food supermarkets for which they were originally envisaged, being found in every size and type of store.

But they extend even beyond that. They are used in industries as diverse as pharmaceuticals, food, timber and steel. They are used to automatically control and replenish goods in a variety of environments. They can identify the cow and the farm that provided your meat today and the fishing area and boat that provided your fish yesterday. They are involved in over five billion transactions worldwide every day, a truly astonishing figure. That is quite a development from the cigarette packet tossed across the scanner a few decades ago.

That is quite a story! Well, this is that story.

*****

The story is about <u>bar codes</u>. It is about who first thought of the idea, how it occurred to him and how that original idea changed to become the 'lines' you see today. It will show you how you can 'read' the code and it will tell you how it is used today. But it will also tell you some of the unusual things that have happened in the world of bar codes. It will tell of people who have used it as an art form, others who have collected as many different ones as they could and even people who have linked bar codes to the end of the world.

But the story is also about <u>item numbers</u> – the numbers printed below the bar codes. Take a product with a bar code from your cupboard now and look at the number below the bar code. Did you know that no other product has the same number in your cupboard ... or in your house ... or in your town ... or in your country ... or in the world? How has such a global system been developed, and how can it be managed and controlled?

It is an unending story which will include some predictions on what you can expect to see in the future. Will the bar code last for ever? Of course not. But, if not, what will we see instead and when will we see it?

Above all it is a story about a group of people who had a vision of a different way of working, who had an ongoing passion to achieve it, and who had the dedication and stamina to make it a reality. This largely unrecognised and unheralded group of people came, and still come, from countries and cultures all over the world, and in a world which is often breaking apart, they have worked and argued together to achieve compromise and agreement.

Therefore this is neither a technical book nor a comprehensive history – they can be found elsewhere. It is written simply for those who over the years have asked me what these bar codes are all about. I hope it will not only answer that question but also give you an understanding of what a particular group of people has achieved. And in doing so, I hope you will share in the excitement they have experienced from the results of their work, which have exceeded all their expectations.

PART I - The Old Retail World

1. The Incorruptible Cashier

It all started with a moment of pure inspiration – a Newtonian moment when an apple of thought dropped from a clear blue sky. The place was a Florida beach, the date was January 1948 and the person who was hit by the apple was Joe Woodland.

This moment would lead to a revolution that changed shops and shopping for ever. So it is only appropriate that the first words of this story should honour that moment and that man, Joe Woodland, the father of the barcode. But to share his story and his joy, we must first understand why he was sitting on that beach lost in thought.

And to do that, we will start in a most unlikely place, Ewaso Ngiro in Kenya. It is the summer of 2004.

*****

I am standing in the Sunday market in the small town of Ewaso Ngiro, a few hours drive from Nairobi. The town is half way from the capital to the Masai Mara National Park. It is also situated at the end of a twelve mile length of collapsed road (almost certainly now improved), which is a contender for the most uncomfortable main road in Kenya. Both facts make the town a welcome stopping point. It is late afternoon, hot and dusty, and so different from England, which I left two days ago.

In the small open-air market, goods are laid out all around me on the ground and on trestle tables. There is everything here that the local people might want for their everyday life. It is a scene that can be found every day, not only throughout Kenya but also in communities across the world.

I am about to have my first experience of 'a kill' in Kenya, and I do not need to wait until we reach the game reserve. The predators have been stalking the herd of tourists and have spotted what they have been searching for, a female and her cub that have become separated from the group. The sellers stalk this vulnerable pair of potential customers, my wife and our seventeen year old daughter. It is no contest. Too late they realise their predicament and make only a half-hearted attempt to escape. For in truth they appear to recognise their fate and even seem to welcome it. Not all of these human predators will feed on the carcases but one will be very satisfied. I look across, see my daughter with a blanket wrapped around her shoulders and realise that it is time for my part in this age-old drama to begin.

I am called across to pay for the purchases made from the lucky stallholder. But there is a role to play before this can be done to everyone's satisfaction. My wife tells me that she has negotiated to buy four Masai blankets for the price of two. This information suggests a series of questions which provide predictable answers.

Why do we need a blanket? It might be cold at night when we are camping on safari and we have not brought one. 'But why do we need four? They will come in handy for presents (to unspecified individuals) when we get home. But how are we going to find space in our luggage to carry them for the next two weeks and get them home? Don't be so mean - we should be supporting people by buying their goods.'

My role is almost complete to everyone's satisfaction so I reach for my wallet. The process of paying is so simple, speedy and almost painless. I find a few notes in the wallet and pass them to the stallholder. He puts them into a money bag secured around his waist, takes out a few coins from the bag and gives them to me. I put the coins into my pocket. Everybody is happy. So isn't retailing simple?

Around the world, communities moved long ago from a self-sufficient barter economy to this form of transferring goods using some means of exchange.

Such transfers continue to take place in this way in local markets from Ewaso Ngiro to my home town of Wirksworth in Derbyshire. It has served us for thousands of years and continues to do so today.

But what has this story and these transactions to do with bar codes, Joe Woodland and his flash of inspiration? If you will bear with me a little longer, I will answer that after introducing you to two men for whom retailing was far from simple, James Jacob Ritty and John Henry Patterson. For them, retailing was a nightmare – because they were losing money.

*****

James Ritty was the owner of 'The Empire', a bar-restaurant on South Main Street, Dayton, Ohio in the 1870s. He was well-known and well turned out, sporting a silk hat in and out of his bar. He also had a reputation as an inventor, with several patents for machines in his name. A feature of his establishment was the rotating palm leaf fan on all the tables of his restaurant, devised for the comfort of his patrons, powered by piped water operating motors. His business was popular, flourishing …. and loss-making. And he knew why.

Like other retail establishments which employed staff, he could not handle cash using the 'money bag method' of Ewaso Ngiro. He had also rejected the use of 'the shoe box method' of an open container behind the counter into which all staff put cash and took change for customers. Instead, he had invested in the Common Cash Drawer, an item which offered two particular advantages. Firstly, whereas the 'shoe box' mixed up all notes and coins, the cash drawer had separate compartments for each one, which made the process of giving change much quicker. Secondly, a thief might snatch a 'shoe box' with all the cash takings, but the weight of the cash drawer made it a much more difficult object to remove from the store. So the cash drawer gave him both better customer service and better security. But, despite its advantages, he knew that it was the cause of his problems.

He was losing money by theft, not from his customers but from his staff. He had already sacked some of his bartenders for removing money from the cash drawer, a practise which has come to be known as 'having your hands in the till'. And he had also found instances where bartenders had either charged their friends less than the set price or charged them nothing at all, a practise which is referred to as 'sweethearting'. He knew he was losing money. He knew how he was losing money. But that did not help either to overcome the problem or to prevent him from suffering a nervous breakdown.

It was on a recovery journey to Europe that he glimpsed a solution. During the trip, his interest in all things mechanical led him to note an automatic device that recorded the revolutions of the ship's propeller shaft. With an inventor's imagination, Ritty related this process to his store in Dayton. Surely, he thought, if a device could count movements of a propeller on a ship, one could also be created that could count movements of cash in his store.

On returning home, James with his brother John, set about making such a machine. Four prototypes were developed before he was satisfied that he had produced something that addressed both of his staff security problems. His earliest machines tried to address the issue of sweethearting. Keys were pressed to record the value of a sale and the keys were connected to a mechanism that displayed this value. Though the first machine showed this value by a clock face with hands, the later versions gave figures printed on tablets which popped-up above the keys. This indication of a sale was still seen on machines in use a hundred years later. Ritty argued that if anybody in a store, including the owner, could see the value that had been charged by a bartender, it would discourage employees from undercharging and sweethearting. And by adding a bell that rang when the keys were depressed, the owner could check that the customer had been charged and that the other part of sweethearting (giving goods away for nothing) was not happening.

The fourth prototype addressed the other security issue of 'hands in the till'. As each key was depressed, a pin pricked a hole in a roll of paper.

At the end of the day, the owner could count the number of holes made for each value of key and thus calculate how much money should be in the till. This sum could be compared with the actual cash in the till to determine if any had been removed.

Although we shall honour Joe Woodland shortly for his contribution to retailing, the equally unsung James Ritty merits early mention in this story for his invention. His creative skills extended to his choice of a name for his new machine, wonderfully called 'Ritty's Incorruptible Cashier'. Unfortunately he did not gain the financial rewards that his efforts deserved. It is often the case that one individual is unable to combine the insight of the inventor and the ruthlessness of the businessman. So whilst Ritty earns our gratitude for his inventive mind, it is a different individual, one John Henry Patterson who will now be acknowledged for making his fortune by developing these ideas. To meet him, we do not have to travel far.

Ritty's Incorruptible Cashier, 1879

*****

John H. Patterson (image courtesy of the engineers Club of Dayton)

James Ritty, inventor of the Cash Register, 1879

John Patterson and James Ritty had much in common. Both lived in Dayton Ohio, both were retailers, both were successful in many ways, but both of their businesses lost money. John Patterson, with his brother Frank, owned the Miners Supply Co, a store that sold coal and general supplies. His problems were similar to those of Ritty.

This is what he wrote about his experience :

'We were obliged to be away from the store most of the time so we employed a superintendent. At the end of three years, although we sold annually about $50,000 of goods on which there was a large margin, we found ourselves worse off than nothing.

We were in debt and we could not account for it, because we lost nothing by bad debt and no goods had been stolen. But one day I found several bread tickets lying around loose and discovered that our oldest clerk was favouring his friends by selling below the regular prices. Another day I noticed a certain credit customer buying groceries. At night, on looking over the blotter, I found that the clerk had forgotten to make any entry of it. This set me thinking that the goods might often go out of the store in this way, without our ever getting a cent for them.'

The Patterson brothers were the victims of sweethearting. But, at this point, the brothers saw a leaflet promoting the Incorruptible Cashier. The product was being sold by the National Manufacturing Company in Dayton, the cash register business and patents having been sold by Ritty to Jacob Eckert for $1,000 in 1881. Patterson wrote later :

'The price was $100. We telegraphed for two of them and when we saw them we were astonished at the cost. They were made mostly of wood, had no cash drawer and were very crude. But we put them in the store.'

The effect of the machines was felt immediately. Within six months their debt had been reduced from $16,000 to $3,000 and the retail store was making profits instead of losses. But, instead of being satisfied with this development, John and Frank Patterson thought about the change that the Incorruptible Cashier had made to their business. They decided that, whilst they could now make a good living from the retail business, they could make a fortune from selling these new registers.

So, they approached Jacob Eckert in 1884, bought the business and patents for $6,500 and changed the name to The National Cash Register Company (NCR). Eckert had good reason to be well satisfied with the deal, for not only had he made a profit of 550% from the sale after only three years, but he also knew that, like Ritty before him, he was running his business at a loss. The Pattersons soon found out too. They went back to Eckert and offered to pay him $2,000 to get out of the deal but were told by Eckert, '*I would not have it back as a gift'*.

The pattern of history is strange, depending as it does so often on the fickleness of random, unplanned events. For at this point John Patterson was saddled with a cash register business which was making a loss and which he did not want. But he had to make the best of it. And he made his fortune.

Ritty was to say later in life, '*I simply could not conduct the cash register business. If anyone other than John H. Patterson had gotten the business, the cash register industry would never have been a success*'. But, as we shall see, Patterson was a very determined man.

The National Cash Register Company (NCR) was set up to manufacture and sell the new register. Its success was immediate and was based on Patterson's personal experience and ability to market the defects of the cash drawer and the benefits of the cash register. This is how he expressed himself in an early sales leaflet :

I am the oldest criminal in history.
I have acted in my present capacity for many thousands of years.
I have been entrusted with millions of dollars.
I have lost a great deal of this money.
I have constantly held temptation before those who have come into contact with me.
I have placed a burden upon the strong, and broken down the weak.
I have caused the downfall of many honest and ambitious young people.
I have ruined many business men who deserved success.

I have betrayed the trust of those who have depended upon me.
I am a thing of the past, a dead issue.
I am a failure.
I am the OPEN CASH DRAWER.

This leaflet is only one example of the marketing and general business flair which Patterson now showed. The success of the business was astonishing. In the period from 1884 to 1914 NCR sold more than 1.5 million cash registers. By 1912 it was selling 95% of all cash registers in the U.S. This success was achieved for a variety of reasons.

First and foremost, as we have seen, the product addressed security problems shared by all retailers and was an outstanding capital investment for them. Secondly, NCR invested heavily in improving its products ; the separate cash register and cash drawer were combined into one unit (one model had nine cash drawers to be used by nine sales assistants), additional features such as records of sales by each sales assistant and customer counts were offered, and the quality and reliability of cash registers were improved. During this period the cast brass-encased register, still found in antique markets today, was the most popular product, so much so that NCR ran the largest brass foundry in the world.

Thirdly, Patterson used new marketing techniques. He is regarded by many as the father of American salesmanship. Although not all of his ideas were original, he introduced sales territories, sales conferences, direct mail shots and formal selling techniques. His booklet 'How I sell National Cash Registers', referred to as the Primer, is still quoted today. Salesmen were told to learn it, follow it word for word, and were fired if they did not do so. But Patterson was not only systematic and disciplined in selling, he was ruthless in eliminating almost all competitors. This was done by aggressive selling techniques used on customers who were considering other products, buying some of the businesses, and legal actions on key competitors. The story is told of the Heintz Cash Register Company which made a machine that replaced the cash register bell with a bird that would 'cuckoo'.

But NCR's challenge on patent infringement eliminated the company and ensured that the last cuckoo had been heard. As a result retail stores for almost a hundred years were to ring with the sound of bells rather than resound to the song of the birds. The effect of all these actions was to reduce the number of competitors from almost 100 to only a handful. A further effect was that Patterson and 29 employees were sentenced to one year in Miami County Jail for Sherman Anti-Trust and restraint of trade violations, though this sentence was later withdrawn before it had been served.

If Patterson was autocratic and ruthless, he was also benevolent. He introduced many social welfare programs for his staff, was involved in much charitable activity, and contributed substantially to the World War I effort. He died in 1922 but had given away much of his fortune by then.

*****

At last it is time to join Joe Woodland and the bar code. But why have I spent so long telling the stories of Ritty and Patterson? A number of themes will run through this book relating to general problems experienced by shopkeepers and attempts to find solutions to them. One of these themes is 'the checkout' – how can you deal with customers efficiently at the point where they pay for their purchases. Another is 'employee dishonesty' – how can you create an environment which will make such theft both more difficult to achieve and easier to detect.

The introduction and ongoing improvements to the cash register made a contribution to these issues. To a limited extent, it reduced the time taken to calculate the total purchase and handle payment. To a somewhat greater extent, it reduced staff theft. However, it was not a panacea.

Staff theft was not wholly eliminated and it is interesting to note that many retailers, particularly department stores, had taken a different approach to the subject by moving the change-giving process off the shop floor.

Compressed air was used to send tubes with money from the sales floor to an office and change was returned by the tube. This widespread system had been in operation since the 1880s and, though it reduced losses from employee theft, it did so at the expense of much slower customer service.

But in busy stores, whilst the cash register had been a major advance, some retailers searched for another invention which would address these problems in a far more significant way.

And so his moment has arrived - enter Joe Woodland.

2. Bull's Eye !

Samuel Nathan Friedland hardly merits any mention in this book. In the late 1920s he opened the Reading Giant Price Cutter store in Harrisburg, Pennsylvania. Its success led to the opening of more stores and the purchase of other chains under the Food Fair name. He only appears here because he inadvertently started the process which led to the development of the bar code.

He was not satisfied with the world of the cash register. In particular, he knew that the process of serving customers at the checkouts using cash registers in his stores was too slow. He knew that if he could serve customers more quickly, he could not only reduce their complaints but also cut the number of checkout operators he employed. Surely, he argued, there must be a better way. There must be a technical solution to speed this part of retailing. So he approached the Drexel Institute of Technology in Philadelphia in 1947 with an idea for a research project. He focussed on the fact that checkout operators had to read the price of every product from its packaging or price ticket and key enter each price into the cash registers, a process which was slow and, for him, too expensive. He wanted to find a method of automatically reading the price of an item directly from the product itself into the registers. If this could be achieved, it would be as big a step forward as was the development of the cash register itself. It would not only speed up the process, but also reduce sweethearting because the operator would only be able to charge the price read from the product, the full price. The Dean of the Institute was not interested – and so went his opportunity for a place in history as the man who recognised the significance of the problem and funded the research which led to the most important retail event of the twentieth century.

However, one of his graduate students overheard the conversation and was fascinated by the problem. Bernard Silver was an electrical engineer and he discussed the problem with a recently-qualified Drexel mechanical engineer, Norman Joseph Woodland.

Their specialisms were to prove complementary in finding a solution. Their early part-time efforts during the remainder of 1947 were unsuccessful. However, Joe Woodland became so obsessed by the problem that he cashed in some stock investments, left his job and went to stay in his grandparents' apartment in Miami Beach, Florida to search for a retail automation solution. He was 27 years old.

Inspiration would not come. So one day in January 1948 he went down to the South Beach and sat on a deck chair to relax. But the subject would not go away. And he started again from scratch. What was the problem? He knew that he could not simply print the price as a series of numbers because there was no technology capable of reading the printed numbers. So he had to find an alternative way of representing the price on the product – some sort of code which could be put on a product, read easily at the checkout and translated automatically back into the price. Surely there must be some way to achieve that.

He thought that he and Bob Silver had made some progress back in Philadelphia. The technical answer would be based on shining light onto this 'unknown code' and turning the reflected light into electrical impulses. They had already conducted some experiments on this basis by taking some printing in different colours, so that the different colours gave different reflections and different impulses.

But this had proved unreliable and they had concluded that a simpler and more practical solution would involve only printing the 'unknown code' in black and white. This would mean that light shone on black printing would be absorbed but light shone on the white background would be reflected. So the electrical impulses would distinguish the black and white and 'read' what was printed.

That was OK as far as it went but they were no nearer finding a code that could represent the price in black and white. Woodland also knew that he had a major practical difficulty with any possible solution – product packaging.

Any printed code would only be one small element among all the words and images found on the packaging of any product. When you thought about all the sizes and shapes of products, how could you design a code so that the sales assistant at a retail checkout would find the code quickly and then position it against the light source so that it could be read? If he could not achieve that, any solution could be even slower than simply finding and keying in a price – hardly likely to commend itself to a retailer wanting to save time. Following that argument meant that the code had to be large enough and obvious enough to be seen easily on every product. This implied that the code could not be one-dimensional (in the form of a variable length or broken line) but would need to be two-dimensional (giving reasonable height and width, and visibility).

He thought about it again. What codes did he know? Well, there was the Morse Code he had learnt when he was a teenager in the Boy Scouts. That system gave each letter of the alphabet and each number a unique combination of dots and dashes. For example :

number 0 was	- - - - -	**5 dashes**
number 1 was dashes	. - - - -	**dot followed by 4**
number 9 was a dot	- - - - .	**4 dashes followed by etc.**

This code, invented by Samuel Morse for use in telegraphy, had been in use for over a century and had been so successful because it was so simple. But Woodland knew he could not use the Morse Code to solve his problem, because although numbers (and thus prices) could be printed in this way, it was one-dimensional and the checkout operator would never find the set of small dots and dashes on the packaging.

Then he had his 'Newtonian moment'. This is what he said about it some years later :
'What I'm going to tell you sounds like a fairy tale. I remember I was thinking about dots and dashes when I poked my four fingers into the sand and for whatever reason – I didn't know – I pulled my hand toward me and drew four lines. I said, 'Golly!

Now I have four lines and they could be wide lines and narrow lines instead of dots and dashes. Now I have a better chance of finding the doggone thing'. Then only seconds later, I took my fingers – they were still in the sand – and I swept them round into a full circle.'

It sounds so simple – so why hadn't he found it before? Well, have you never looked at a crossword puzzle and been completely stumped by a particular clue, but then returned to it later and seen the answer immediately? How can you explain that? And is there anything more frustrating than to be told that 'the answer's simple'?

Murder has been committed for less. There is no such thing as a simple answer.

And the fact that this solution is so clear and neat does not detract from the magic and significance of what Joe Woodland had found.

*****

Woodland had solved two linked problems. The first was how to create a code that could be read as a number (i.e. the price of a product). The second was how to make it a practical code that could be easily found on the packaging of products and read at the checkout.

Let's consider how he solved the first of these. He took as his basis the binary number system where 001 was one, 010 was two, 011 was three, 100 was four ..... up to 111 which was seven. If he printed lines instead of ones and spaces instead of zeros, then 100 could be printed as 'line, no line, no line' and translated as the number four. For a three digit binary number it meant that he could represent up to the number 7 by lines. And he could add more lines. So a four line print would be 15 (or 15 cents), seven lines would be as much as 127 (or $1.27) and twelve lines 2047 (or $20.47).

This is simply an illustration of how a series of lines <u>could</u> be read by a light source and, by seeing which lines were printed, converted into a number or price.

But Joe Woodland did better than this illustration. As he said at the time, he could make the lines of different thickness. By varying the thickness of lines and distance between them, he could translate the line code into a number using fewer lines. At this stage we will only note that it is possible and return to his idea later.

But there are still difficulties with our series of lines. How can you tell where the set of lines and spaces begins, because the first 'digit' may be a line or a space? Woodland knew that he could solve this problem by printing a recognisable 'identification line' before the code. But then what happens if the checkout operator does not position the scanner on the product correctly and reads the code upside down? The sequence of lines and spaces (and the price) in that case would be different. Or suppose the positioning is such that the lines appear to the scanner to be horizontal rather than vertical making the code unreadable.

Clearly we cannot ask the operator to place each product under the reader with great care – the resulting checkout queues would be even longer. But that is where Woodland solved the second problem by thinking not of parallel straight lines but of a circle of lines. If you think of those lines, providing the scanner reads the centre of the circle, it does not matter which way you hold and read the code, you will always get the same result. So you could say that Woodland had hit the bull's eye with his thinking – he had created a Bull's Eye Code.

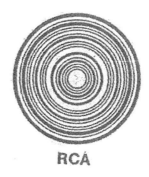

RCA

29

Woodland and Silver developed the idea further and applied for a US Patent for their invention on 20th October 1949. It was titled 'Classifying Apparatus and Method' and described as relating to 'the art of classifying items based on photo-response to a set pattern of lines'.

Although Woodland's thoughts on the beach had led him to the bull's eye, the actual patent application consisted of a pattern of straight lines with four white lines on a dark background, read as 0 to 7. On 7th October 1952 US Patent No. 2612994 was finally issued and the bar code revolution had begun.

*****

At this stage, it is possible that you have a number of questions. As you look at the bar code printed on a product and consider the checkout operation in a supermarket today, you may be puzzled about some of the things I have said. For instance :

- If the Bull's Eye Code was such a clever idea and the Straight Line Code was such a bad idea, why are all the bar codes on products today a series of straight lines with not a single bull's eye in sight?
- If reading a set of straight lines from different directions was a problem, how do scanners today seem to read the bar code however it is presented by checkout operators?
- If you can print a number or price using only a few thin or thick lines, why do most of the bar codes on products today have 30 lines? Why do they need as many lines as that?
- Underneath the bar code on products there is a printed number. Why is that necessary and what does that number mean?
- With such a lot of lines and such a long number, where is the price? It clearly can't be all those lines so is it some of the lines or part of the number or both?

These questions arise because we are only part way through the story of the bar code. And if it seems that the subject is becoming more complicated, can I seek your patience? All of these questions will be answered in later chapters when a simple explanation of today's bar codes will be given. In understanding that, you will be able to recount the tale of incorruptible cashiers, cash registers and bull's eye codes to explain to others why today's bar code is so much better.

But I must now conclude this part of the story by following the lives of Woodland, Silver and the Bull's Eye Code. In 1952, their ideas were of limited practical use. They required much more technically advanced devices to read the bar code and low power laser technology was still twenty years away.

They also required much more concentrated computing power to interpret what was read and these were the days when a computer with a fraction of the power and memory of your personal computer required a room full of equipment. Despite this, they continued their development work. In 1958 they secured a further US patent and this time it was for a Bull's Eye Label. Bernard Silver died only four years later in 1962 at the age of 38. He never saw the commercial use of the bar code. Joseph Woodland continued the work they had started, saw its astonishing life and was acclaimed for his role.

In 1992 President George Bush Senior presented a small number of recipients with a National Medal of Technology. One of them was a 37 year old who is quite well known and another a less well known 72 year old :

'For his early vision of universal computing at home and in the office ; for his technical and business management skills in creating a worldwide technology company ; and for his contribution to the development of the personal computer industry' , the medal was awarded to William H Gates III (Bill Gates) of Microsoft Corp.

'For his invention and contribution to the commercialization of bar code technology which improved productivity in every industrial sector and gave rise to the bar code industry', the medal was awarded to Norman Joseph Woodland of IBM Corp.

But what about the Bull's Eye Code? Well, it was developed further, it was used in supermarkets and it did experience a short period of fame. In the 1950s Woodland joined IBM. Woodland and Silver tried to sell IBM their bull's eye design but, when that company was not prepared to pay their asking price, they sold it to the Philco Corporation. Philco in turn sold it to RCA, who were involved in its commercial development. Finally, in July 1972, twenty years after the first bar code patent, the first experimental fully automated supermarket was opened. Kroger's Kenwood store near Cincinnati used Bull's Eye Codes on products throughout the store, some printed on the packaging but most with stuck-on labels. The Kroger brochure celebrated the event :

'It takes two to tango and it takes two to make food shopping better ... YOU ... and your SUPERMARKET! With the cooperation of RCA Corporation, whose electronic wizards created the new scanner checkout now being tested at the KROGER KENWOOD PLAZA STORE, Kroger is taking the first step toward bringing your fondest food shopping dreams into reality!'

A prototype of the checkout stand used in this Kroger test can still be seen at the Smithsonian National Museum of American History in Washington.

But the bull's eye code had a limited life. In 1971, shortly before the Kroger test, an RCA demonstration of its code at a grocery industry meeting had been so popular that IBM managers present realised that they needed to put more effort into this market. They transferred one of their members of staff to North Carolina to work on an IBM alternative. This employee would help to kill the bull's eye code and ensure that the bar code world became one of straight lines rather than circles.

With a delicious irony the man they chose to achieve this was .....
Joe Woodland.

******

Our story so far has been about retail problems at the checkout
(speed of customer service, increasing staff costs and cash security
reducing profits). It has taken place exclusively in the United
States and has introduced the subject of bar codes. We shall return
shortly to the US, to bar codes and to Joe Woodland.
But before doing so, we need to hear about another set of events
taking place at this time thousands of miles away.

For the next part of our story is about a different set of retail
problems. It takes place in Europe and will introduce the subject
of article numbers. It is time to introduce Bundes Artikelnummer
Lebensmittel (Ban-L) and Groupement d'Etude de Normalisation
et de Codification (Gencod).

3. BAN-L and GENCOD

The story so far has been exclusively concerned with the visible part of retailing, the part you see as a customer at the checkout. Your own experience will have helped your understanding of why retailers sought solutions to the problems of slow customer service, sweethearting and 'hands in the till'.

We now move to a part of the retail world that you will probably not have visited and to a retail operation that you have not experienced. In the backrooms of shops and the Head Offices of retail chains, there was another subject which gave great concern. The problem is simple to explain. It is about bits of paper, millions and millions of bits of paper.

To understand it, we will start by going to Hamburg in 1960 and visiting the offices of Edeka, a German retail food chain. There is paper everywhere. Most of it relates to the process of getting a product on to a shelf in a store, the bit the customer never sees. Consider the number of pieces of paper and actions that are needed for this to take place :

- Product Lists from manufacturers with the items they are trying to sell to Edeka and their prices.
- Order forms from Edeka with selected products, quantities and prices to be sent to the manufacturers.
- Delivery notes from manufacturers with products and quantities sent to Edeka (which have been checked against the actual goods received and any discrepancies noted).
- Invoices from manufacturers with products, quantities and prices charged (which have been checked against the original order and the delivery note to ensure that the actual goods ordered and delivered have been charged at the correct price).
- Cheques from Edeka for agreed payment to be sent to the manufacturer.
- Statements from manufacturers detailing invoices issued, payments made and the balance owed by Edeka.

By the 1960s, Edeka had used early computers to automate part of this process. To do this the company had devised an internal product coding system using numbers and/or letters for all of the products they sold in the stores. At the same time, a few manufacturers were also using computers to automate their processes and they had devised their internal product coding systems for all the products they sold to retailers. Needless to say, each retailer and each manufacturer had their own unique coding systems and every transaction required a conversion from the manufacturer code to the retailer code.

But as Karlheinz Hagen, who would later be heavily involved in the German bar code revolution, remembers :

'Edeka was a retail food chain. At their headquarters in Hamburg, Edeka staff prepared manufacturers' invoices for payment. They took an invoice with the manufacturer's product numbers, looked for the equivalent Edeka item numbers listed in a catalogue, wrote these numbers on the invoice with a red pen, and sent the invoice to a punch card operator who used the red Edeka numbers to enter the invoice details onto a punch card for computer input. Of course, one person in Edeka had a great idea – ask the manufacturers to put the Edeka numbers onto the invoice and save us work. The manufacturers were not happy! They thought about the number of different retail and wholesale chains, all with their own different numbering systems, and thought that such a change would set a precedent.'

There seemed no easy way out of the problem. No manufacturer wanted to convert their own product codes into the different codes of each retailer. Equally no retailer wanted to convert their codes into manufacturer codes. However, Karl-Heinz Severing summarised the opportunity and suggested a way forward to a retail industry committee in 1966 :

'The food trade has just started to introduce electronic data processing. Perhaps 1000 companies could use such systems but fewer than 50 have been installed so far.

With EDP, retailers and wholesalers are just a step away from integrated data processing with their suppliers. For example, it is possible to process automatically all incoming invoices without manual processes. It is conceivable that the supplier could provide a 'punched card' instead of a printed invoice which would contain all the invoice details and could be entered directly into the computer. But to allow the industry to adopt these new technical possibilities, it is necessary for everyone involved to use the same item number and supplier number keys. The big question is whether we can all agree on a unified item number and supplier number system.'

So there was the simple solution, a sort of Esperanto in numbers for items and companies. But, wait a minute. Surely such a system would mean that instead of either retailers or manufacturers learning the other's 'language of numbers', both would have to learn a completely new language? Yet that is precisely what was proposed and what the industry agreed to do.

Why did they do so and why was it so successful? Perhaps it was a satisfactory political compromise – both sides of the industry would share the pain and cost of change. Perhaps it was the right time for change – many companies were investing in new systems and this could be accommodated at modest cost with clear benefits. Perhaps it was because they were German – with a reputation for efficiency, discipline and working together. And perhaps it was because of leadership – for at this stage the key promoters of the project got the powerful support and sponsorship of the Federal Ministry of Economics, the Bundeswirtschaftsministerium.

*****

Later in our story, we will look at the Article Number system which you see today on almost every product you buy – how it is structured and how it is managed. Before that, this first major national Article Number system in Europe will be summarised because it is fundamental to understanding today's very different international system.

The Germans decided that they would give every existing item that was traded between a grocery manufacturer and retailer a unique number. However, whereas the purchases made in a food store by customers are generally for single products, those products bought by the store from manufacturers were delivered to them in larger packs (or outers). And because the numbering system was only designed to improve the administration between manufacturers and retailers (which did not involve customers in the store), numbers were only given to these outer packs of products bought by retailers from manufacturers, not the single products bought by customers from retailers.

Thus for example, a jar of jam such as Ananas-Konfiture Portionsgl 50gm, sold in a retail store was not given a code. As a result, retail customers were unaware that any number system existed, because no number was printed on any item in the store or visible anywhere within the store.

However, when that jar of jam was part of a pack of 30 jars sold by the manufacturer, the outer pack had the code 05603787. And when it was part of a different pack of 180 jars and sold at a different price, the pack was given a different code 05603794. Not only did the Germans decide that each traded product meant giving different numbers to different pack sizes, they also decided that different flavours of the same product needed different numbers. For example :

Eka Gold-Konfiture Kirsch 450gm (pack of 12) had a code
    05601668
Eka Gold-Konfiture Aprikose 450gm (pack of 12) had a different code   05601675

But how was each different pack of products allocated its unique number? Well, the Germans could have given the first number to be brought for coding the number 1, the second product number 2 etc. However, they agreed that they needed a much more structured system before any numbers were allocated. So the German grocery industry was analysed in discussions which involved 80 different associations of manufacturers.

As a result it was decided that all the numbered products should be separated into 100 Product Categories, that each of these Categories would be sub-divided into 10 Product Groups, and that each of these 10 Product Groups would be sub-divided in turn into 10 Article Groups. In other words, any item which was presented for coding had first to be allocated to an Article Group within a Product Group within a Product Category (i.e. into one of 10,000 classes of products)!

Without wishing to stereotype, this may seem to some to be a very German way of doing things. It is very structured, needs detailed organisation to maintain it and great discipline. There is no doubt that a distinctive German way of thinking was significant in its design, as can be seen by the fact that other systems at this time did not follow this pattern. But there was another consideration which led to the decision to have such a system.

A key part in the design of the system was a printed catalogue listing all the products that had been allocated numbers. In order for that to be readable, it was necessary to print it in sections of different types of product.

So the numbering structure meant that the numbers could be printed in ascending sequence and they would automatically appear in 100 Product Categories. And the pages for each Product Category were printed in Product Group sequence with sub-sections in Article Group sequence.

And so the Bundes Artikelnummer Lebensmittel (Grocery Article Number System) was born. The BAN-L system involved articles which were all allocated 8 digit numbers with the following structure :

Pc Pc Pg Ag Pn Pn Pn Ch where

Pc was 00 to 99 (one of 100 Product Categories)
Pg was 0 to 9 (one of 10 Product Groups)
Ag was 0 to 9 (one of 10 Article Groups)
Pn was 000 to 999 (a Product Number)
Ch was a check digit

A central organisation was established to allocate Article Numbers and ensure that every new product was given a unique number in the appropriate product category, product group and article group. Karlheinz Hagen remembers its launch on 1st September 1969 :

'The Ban-L system involved a printed catalogue of products. Manufacturers realised that each Class would have many pages of Product Numbers and that retailers would need to read these pages to find particular products. They thought it was important that their own numbers were always seen by being at the top of the list. So they needed to apply early for their own numbers and secure the 'low numbers'. When the doors opened at midnight there were hundreds of companies waiting with their list of products!'

The Ban-L system was far more detailed than the description I have given. But you will be grateful for the fact that we can ignore the remaining large range of rigidly defined elements that allowed manufacturers and retailers to exchange information automatically. Our story is only concerned with Article Numbers and the impact made by this first major industry-wide system in a European country. It was immensely successful. Not only had over 60,000 article numbers been issued in Germany by 1976 but also a separate Ban-Austria system had been introduced in 1972 using the same fundamental number and organisation structure.

However, our very limited archaeological dig of numbering systems has unearthed a dinosaur. For, as we shall soon see, no trace of this system now exists. Perfectly developed for its environment, it would shortly be hit by an external force that would consign it to history.

But to complete our understanding of early Article Number systems, we must make one further visit – to Paris. In France, manufacturers and retailers had identified the same problem of different coding systems as in Germany and they saw the same opportunity for a national system.

So they would have reached the same conclusions and come to the same solution as the Germans, wouldn't they?

*****

The detailed, highly organised system that was developed and implemented by the grocery industry in Germany did not have the same appeal to, how can I put it, the somewhat freer spirits of the French. In that case, you might conclude that they would be searching for a solution that was somewhat more ..... French. And you would be right. But we must be careful not to bring to this subject our own preconceptions, for there were very practical reasons why the French conclusion was almost the opposite to that of the Germans. In particular, the starting point in each country was not the same.

In France, as in Germany, many manufacturers and many retailers had developed their own numbering systems. And again, ever-more powerful retailers were beginning to demand that their own numbers must be used in communications. So, manufacturers were faced with the horrifying prospect, as in Germany, of referring to any one of their own products by a different number for each retail chain.

However, in France, interest in the subject was not simply from grocery chains. General merchandise chains also wanted a national numbering system and so the number of products involved became much greater. Furthermore a significant number of these products were from the textile industry, where it was made clear that in their business every size of every colour of every product needed a different number, leading to greater complexity and a much larger number bank.

A further key difference to the German situation was that one significant retail chain had already implemented a system which had led them to give every one of their manufacturers or suppliers a unique supplier number.

But not only was the starting point different, the approach to the subject was distinctive. The French did not want a centralised bureaucratic system or a national catalogue of numbers. Instead they looked for a 'light touch' approach which could still give unique numbers to products. Their solution was a simple one. They would set up a small central organisation that would only be responsible for giving each supplier a unique supplier number of fixed length. Each supplier would then have the responsibility for numbering their own products, again with a fixed number of digits.

The advantages of that approach were that central costs would be much less, numbering products would be much faster, and many suppliers could continue to use their own existing number systems. The disadvantages would be that it would result in a much longer number, numbers would be allocated from many different places, and there was no central knowledge of all the numbers created (and thus no realistic possibility of creating any future national catalogue). However, the French decided that they could live with a longer number, that their number structure would limit the possibility of any duplication of numbers, and they would never need a national catalogue.

So the French 11-digit article number was created as :
Co Co Co Co Co P P P P P P where

Co was the 5-digit Company Number allocated centrally
P was the 6-digit Product Number allocated within each company

The 6-digit Product Number could be used within a company for internal communication but the 11-digit Article Number (which identified the company and the company's product) must be used for inter-company communication.

*****

So in summary, by the early 1970s, there was a system for communication in both Germany and Austria between manufacturers and retailers based on unique 8-digit Article Numbers. Numbers had a detailed four-part structure and were allocated by a central numbering authority. In France there was a different 11-digit Article Number system. There numbers were split only into two parts, with the company number allocated centrally and the internal item number added by each company. In both countries these systems represented a major step forward in improving communications between manufacturers and retailers, as well as reducing the associated costs.

Two different but equally satisfactory solutions. However, it does lead to one rather awkward question. These systems apply to trading relations within either Germany and Austria or France. But what if companies want to trade across national boundaries? The numbering systems are completely different, so which system should they use for international trade, the German one or the French one? Well er ….. that was a major problem and the subject of much discussion. It is a subject which we will come back to shortly. But before doing so, we must go back across the Atlantic Ocean and take further the progress of bar codes and Joe Woodland.

4. Coming Together

It is at this point that the two paths we have followed so far (the bar code in the United States and the article number in Europe) merge and become a lane that will later form a road and finally a motorway. For the subjects of the bar code and article numbering are not separate but totally locked together. However the journey along those roads was not entirely smooth.

In the US a retail industry meeting held in Cincinnati in September 1969 was a disaster. On one side of the table were systems representatives from the major manufacturers of branded products sold in grocery stores in the United States. On the other side were systems representatives of the major supermarket chains. Each side had its own agenda. For the retailers, the major issue was simply stated – how could they reduce staff costs at checkouts. This subject was the same one that had been raised in 1947 and which had led Joe Woodland to look for a technical solution. In these 22 years the issue had become more acute because staff costs had risen much faster than other retailing costs and no technical solution had been implemented. By 1969, the largest supermarkets found that, after taking the actual costs of the products they bought, the total costs of labour represented two-thirds of their remaining expenditure. And much of that was for checkout staff. So they wanted to speed up the checkout process and reduce the number of staff required.

But they were looking for something different to the 'price code' that Woodland had searched for earlier. Like Germany and France, the solution they envisaged was to introduce an inter-industry product number/code for all products, but with the key addition that it could be converted into some form of machine-readable symbol which could be read at checkouts. They added two requirements – the number/code must be short (seven digits) to reduce the cost of computer storage capacity in holding all the numbers, and it must have no detailed meaning built into it (in their technical language the number must be 'non-significant').

For the manufacturers, the issues were more complex. They could see some advantages from a standard product code. However, each of them had their own existing independent coding systems which were generally of four, five or six digits and they were adamant that any inter-industry code must incorporate their own internal codes. In that sense they were following the same logic that was being pursued in France. As a result, they also added two requirements – the code must be long (eleven digits) and it must have meaningful information built into it (i.e. be 'significant'). Not only did they have a problem with codes, they also had difficulties with symbols. They were worried that even if a uniform code was agreed, the code might be converted into different symbols by different individual retailers or groups of retailers. In that situation, they might be forced to print all of the symbols on their packaging to avoid a legal anti-trust challenge stating that they were giving some retailers preferential treatment. So they were keen to find one symbol which was common to all.

Both parties wanted a solution but their requirements were incompatible. Stephen Brown, the legal counsel who was not only involved at that time but also for many years to come, has written about the meeting's outcome in his history, 'Revolution at the Checkout Counter' :

'As had happened in the past, one group said, 'eleven digits, significant', the other responded 'seven digits, non-significant'. After staring at each other for a time, the groups adjourned the meeting and went their separate ways.'

*****

How do you resolve an issue like this? There was no lack of desire from those involved. But their requirements were different and, try as they might, they were unable to find a mutually satisfactory solution.

The way the deadlock was broken is worthy of inclusion in business, political and social science textbooks.

Clarence Adamy, of the National Association of Food Chains, concluded that the group of systems staff who had met would never find a solution. He convened a meeting of a number of retailer and manufacturer trade association presidents. They in turn concluded that the problems to be solved were not fundamentally technical, which could be solved by systems managers, but political, which must be solved by business managers. Furthermore they decided that any far-reaching political solution could not be reached by either trade associations or junior and middle business managers. So they set up another meeting, this time for a small group of Chief Executive Officers from key manufacturers and retail chains, arguing that only they were in a position to determine the strategic and financial significance to their companies of finding a solution. Let them decide how much they wanted an answer and what they were prepared to do to get it.

The meeting a year after the failure in Cincinnati was far more successful. It was held in Chicago and chaired by Burt Gookin, the first non-family member to be President of H.J.Heinz. The ten company heads (five retailers/wholesalers and five manufacturers) formed the Ad Hoc Committee for a Uniform Grocery Product Identification Code. And they reached a unanimous agreement immediately. Well, that statement needs some qualification. They agreed that the search for an inter-industry code and symbol was significant for all their businesses. They agreed that a plan was necessary to make it happen. They did not get into any detail on what that code and symbol should be. To decide that, they appointed Tom Wilson and Larry Russell of McKinsey to present an independent cost/benefit assessment and recommendations. As Stephen Brown summarises :

'McKinsey's task was to identify a coding structure that could be used as a standard for item identification by grocery manufacturers, distributors and retailers, yet require the minimum changes to existing coding structures. An economic case would have to be built that the benefits to the industry as a whole, and to each segment of the industry, could justify the expense of making whatever changes were required to existing codes.

A uniform code should also be conducive to a symbolic representation that could lead to the automation of the checkout process. Finally, the ideal code would be structured in such a way as to convey the maximum amount of information in the minimum amount of space.'

That was much easier said than done – but surprisingly (perhaps even to them) they succeeded. A coding structure was agreed by the committee eight months later on 31st March 1971. In that period the committee had met six times.

This involved a substantial time commitment from the chief executives of major companies, but indicated the importance of the subject to them and the pressure that they put on McKinsey to achieve an early solution. So what was their recommendation?

They proposed a numeric product code. That may not seem significant. After all, that is what had been implemented in both Germany and France, that is what we see on products today, and that is part of the title of this book. But it wasn't the only possible conclusion – why didn't they recommend a code consisting of letters?

In 1971, I was working for Boots the Chemists in the United Kingdom. Boots had its own internal product code solution but it was based on letters rather than numbers. And that solution was reached because of one of the requirements that had been stated in McKinsey's brief, *'to convey the maximum amount of information in the minimum amount of space'*. Just think of two different systems based on 10 different digits or 26 different letters for a moment. A two-digit code gives 100 possibilities (10x10) but a two-letter code gives 676 (26x26). A three-digit code gives 1,000 possibilities but a three-letter code gives 17,576. A six-letter sequence gives over 300 times more options than a six-digit number. In other words, for any given coding capacity, a system based on letters would result in a shorter code than one based on numbers. Suppose we want a system with 100 million options. That would require an eight-digit number but could be achieved by only a six-digit alphabetic code.

So, why did McKinsey recommend a code based on numbers which did not minimise the amount of space rather than an alphabetic or mixed alphanumeric code? And why has our world become one of article numbers rather than article letters?

The fundamental reason was contained in the part of the brief which *required the minimum changes to existing coding structures'*. Since very few companies had identified their products using an alphabetic code, it was inevitable that a numeric code would be chosen to minimise the work of conversion. In fact some of those involved at that time have told me that an alphabetic code was never seriously evaluated for this reason alone.

But there were other practical and financial reasons for preferring a numeric code. Keyboards would be required at checkouts to type in a code when the symbol was unreadable. But keyboards to enter only numbers rather than letters were not only smaller and cheaper but also could be used by relatively unskilled staff. In such circumstances, even somewhat longer numeric codes would be typed faster and with fewer errors than shorter alphabetic codes. So a numeric code was chosen and the financial case made by McKinsey for moving ahead was accepted by the committee. But what number was accepted – the retailer's seven-digits non-significant or the manufacturer's eleven-digits significant? As is usually the case when compromise needs to be reached, the answer was – neither.

The agreed new Uniform Grocery Identification Product Code was ten digits, a five-digit company number followed by a five-digit company-allocated product number.

This was remarkably similar to the conclusion reached in France, where the eleven-digit national article number consisted of a five-digit company number and six-digit company-allocated product number. It would be wrong to see this in simple terms as being largely a victory for the group of manufacturers.

It was a soundly-based decision which recognised that, in the short term and particularly in the medium term, a longer number would not increase costs significantly but it would allow the change to the new system to be handled more easily. More than that, it was a decision that remains as the basis of current article numbering forty years later, whether by great thinking, great luck or both.

But if the Committee had moved the subject on a significant step by its work during these eight months and by its final decision, it would be a further two years before a US Product Code (or Number) and Symbol would be fully developed and agreed for release.

*****

In these two years, separate groups working on the Code and the Symbol came up with the solutions we still have today. During this period, the Code changed from the ten-digit starting point and the Symbol, which did not exist at the outset, was developed.

The Code Management Committee increased the length of the code, changed the name of the code and set up a structure for administering the new system. But why change the length of the code when it had taken so much effort to reach agreement on the ten-digit code in the first place? Two issues came to light that resulted in an extra digit being added at the front of the code and an extra digit being added at the back to create a twelve-digit code.

The Committee thought about the work of a checkout operator in a store handling products with the new code and linked symbol. They saw two potential major problems. Firstly, what if the basket of goods included some products which had the new code and symbol but others which were not included in the new grocery coding system? The non-coded, non-grocery items would have to be dealt with in a different way and the checkout operation would be much slower – unless of course the proposed system could be widened to include these products. However, some of those non-grocery products already had numbers which were part of different numbering systems.

Looking at those products, the committee saw a way forward. One digit would be added at the front of the code for different numbering systems. For grocery products this number would be 0 which would be followed by the proposed ten-digit number. For the drug industry the number would be 3 followed by their existing National Drug Code and National Health Related Items Code. For other industries a different number could be allocated and any code up to ten digits could follow it. In this way it was hoped that all products in a store would be part of one wider overall system and all could be handled in the same way.

Secondly, what if the checkout operator keyed in a number incorrectly? Whilst most of the time the Symbol would be read automatically, there would be occasions when the number would need to be key-entered.

And some mistakes were bound to occur.

Where mistakes were made, the most common errors would probably be either keying one digit wrongly (say 4 instead of 3) or keying two consecutive digits the wrong way round (say 97 instead of 79). When millions of customers were being served that would happen and result in the wrong number being entered, possibly for a valid but different product. The committee saw a way forward. An extra digit could be added at the back of the eleven-digit code. This would be a check digit calculated in such a way that most of these types of error would produce an invalid total number. (A full explanation of this calculation is included in Part II.)

Having now changed the length of the code from ten-digits to twelve-digits for these two reasons and widened its potential use beyond grocery products (and possibly even grocery retailers), the committee decided that the name should change too. And so the Uniform Grocery Product Code became the Universal Product Code (UPC), even though the name of the management body remained as the Uniform Code Council (UCC). The change from Uniform to Universal to describe the code is somewhat strange, given that it was only designed to be used in the United States and Canada.

Those involved at the time tell me that there was no discussion or debate about the change, it 'just seemed to happen at one meeting'. But some reading this may feel that it was not the first occasion, and will not be the last, on which a US citizen has considered that there is no world outside the United States!

So we arrive at last with the twelve-digit Universal Product Code :

Px Co Co Co Co Co P P P P P Ch

Px - the Coding System Prefix
Co - Company Number (5)
P - Product Number (5)
Ch - Check Digit

The Code Management Committee had one last proposal to make – how to manage the code. They decided that administrative costs would be minimised by setting up a central body which would allocate company numbers, but leaving each company to add its own product numbers and calculated check digit to create the full UPC for each item. It would then be the responsibility of each company to inform their customer retailers of the codes, with no central database of all products that had been allocated codes by all manufacturers. (This solution was identical to that which was introduced in France.) They authorised an existing organisation, the Distribution Number Bank (DNB), to handle the allocation of company numbers. It is interesting to note that providing numbers which would become the basis of almost every company's computer systems was achieved by a low-tech approach. A set of 100,000 tear-off numbers was printed, and each one was then removed and attached to the membership certificate for a company. The first number was allocated on 15th May 1972.

Half of the job was done. A Universal Product Code had been agreed and established. But what about the linked development? What symbol would represent that code?   That would be the work of the Symbol Selection Committee.

5. Birth of Today's Bar Code

The Dragon is dead. It had a very short life, expiring in 1984 at the age of two. But in that time, the Dragon 32 Computer, with 32kb of RAM, gave my two sons a lot of pleasure and they were very unhappy when I managed to lose it during a house move. Its competitors in the UK, the Sinclair Spectrum and the BBC Micro, had a longer life but have also passed away. The pace of technological change results in a high death rate for products and even companies.

Therefore, unless you are of a certain age and with particular experience, you are unlikely to know RCA, NCR, IBM, Litton, Pitney Bowes and Singer. Yet in the early 1970s they represented leading, well known and well respected companies in the retailing industry who were the finalists in a competition to develop what has become today's bar code. The competition was established, managed and judged by the eight members of the Symbol Selection Committee. It first met on 1st June 1971 and its decision on 3rd April 1973 produced the bar code you see in retail stores today. That symbol, which you see on millions of retail products did not exist when they started their work. In nearly forty years of immense technical and business change, their choice of symbol has been a fixed constant. Furthermore, it is not only still in use for the purposes it was envisaged but it has also been adapted for situations that were either never considered or did not even exist at that time, as we shall see in Parts II and III. It is an awesome achievement and says much for their imagination, energy, methodology and judgement as well as the commitment of the staff from many technology companies. Therefore it is worth going into a little detail about how they went about their task.

The committee was chaired by Alan Haberman, Chief Executive Officer of First National Stores (Finast), a long-established and significant New England supermarket chain. His energy drove this committee forward and remained apparent throughout the following thirty years of his involvement in the bar code world.

He was given a set of operating principles by the Ad Hoc Committee, which emphasised that three groups would need to be satisfied by any conclusion they reached. Two of these were obvious, for both manufacturers and retailers had to reach agreement or the necessary investment from each of them would not take place. But the third group was equally significant - consumers. If supermarket customers did not accept the changes which were proposed, then the project would prove an expensive failure. So the role of customers was accepted from the outset and was to remain central at every stage. The committee had to adhere in their working to a set of general but key points designed to ensure that any final recommendation would be acceptable to all these three groups :

- The proposed Symbol must be selected on the basis of overall industry economics.
- The choice must not place an undue competitive burden on any segment of the total industry.
- All negative points of view must be answered by the committee.
- All requests for information must be honoured.
- Consumer interests must be recognised.

They were also given a series of detailed points regarding any proposed Symbol that could represent the agreed Product Code in a machine-readable format. In total they presented a demanding specification :

- It must have a scanning and reading accuracy of 99.99% (one error in 10,000).
- It must be omni-directional (ie readable from any direction).
- It must be able to be printed on any current and anticipated packaging sizes and materials, and containers.
- It must also be able to be printed within a supermarket at reasonable cost.
- It must have an area no greater than five square inches.

- It must be read in any environmental condition found in a supermarket.
- It must be human-readable as well as machine-readable.

In Part II we will return to each of these points when looking in detail at existing bar codes. For the present, we will simply note that, if this symbol was going to be read billions of times each day on many thousands of different products in many thousands of different stores, and by many different scanners produced by different companies, it would, to say the least, need to be remarkably robust.

The role of the committee was simply to select such a symbol. It was not expected to create possible symbols but only to evaluate the work put forward by others. The development work would be undertaken by technology companies at their own cost. These companies were given a detailed specification and told that, in addition to meeting these requirements, any proposal must include a practical demonstration in an actual supermarket. Oh, and by the way, it was added, the successful symbol specifications must not be kept by the company concerned but made available in the public domain so that any other company could develop products to print and read symbols at reasonable cost.

Why would any company want to commit staff and money to such a project which offered a relatively low chance of success and, even if successful, would be offered immediately to their competitors? The fact that so many companies were interested in development and that six of them presented detailed proposals was an indication of how great they thought the potential total sales and marketing opportunity was going to be. But the six finalists came with very different backgrounds, motivations and expectations.

RCA had to be clear favourites at the outset. They had taken Woodland's Bull's Eye Code and developed it. As a result, half way through the evaluation period of the committee they were able to demonstrate the full working Kroger supermarket system referred to earlier.

They had put significant investment into the project, were clearly ahead of any competition and must have been confident that their efforts were about to be rewarded.

NCR were in a completely different position. They still dominated the cash register market in the United States, as they had done since their founding by John Patterson nearly a century earlier. But that position was under threat. Japanese electronic cash registers had been developed and were increasing their penetration of the US market.
NCR had been forced to invest heavily in their own development efforts and were focussed on fighting this threat to their dominant position. Perhaps because of this, they expressed serious reservations about the whole symbol development project and were very sceptical that it could succeed. Although they were involved, it seems clear that the prospect of a successful symbol, which would open up the supermarket business to another group of competitors, was not appealing to them at a difficult time for their business.

But IBM were licking their lips. To them the symbol represented a massive new business opportunity. The successful introduction of a symbol to be read in supermarkets would require retailers to move from having only cash registers in their stores to needing some form of computerised register. IBM as the leading computer company anticipated that many of these new computers would be bought from them. Strategically it was massively important to them that this project succeeded and they were wholly committed to making sure that it did. As we have seen (in the introduction to this book) they had already done some work on scanning bar codes.

By April 1973, the committee felt able to assess the six solutions presented. In the light of history, the decisions to seek a symbol and to select a particular symbol may seem obvious. They were not as obvious at the time. Behind the decisions lie many disagreements and alternatives that were rejected, which could have produced a different outcome.

For the purposes of this book, it is sufficient to note that the decision to seek a symbol and to ensure that one was agreed was driven by a group of individuals and companies (retailers, manufacturers and equipment suppliers) who needed a positive and speedy outcome. So what did they choose?

*****

U.P.C. Standard Symbol

I have not found any record of the formal reasons why the committee rejected the RCA Bull and accepted the IBM Lines. I am told that the major opposition to the Bull's Eye Code came from the printing industry, who expressed severe reservations about their capability of printing such a code on all types of surface in such a way as to minimise errors in printing and reading. In particular, a printing problem of 'print spread' represented a fundamental issue for reading circular lines but was less significant for straight lines.

Equally there is no formal record of why the committee did not choose simply to accept a machine-readable number with no further symbol. Optical number recognition (OCR) did offer an alternative solution. However, reading numbers required the number to be held in a stationary position above the reader. As such it would have slowed the checkout process to an unacceptable degree.

In addition, whilst the error rate of one misread per 10,000 numbers scanned in this way would have met the minimum accuracy specified, bar code reading gave much better results. (Current laser scanners can give an accuracy of only one misread per 70 million reads.) Finally the committee may have been concerned that a printed number-only solution would be less visible than a bar code on packaging. But, although number-only was rejected, a machine-readable number was part of the final recommendation.

The decision taken was to accept a solution with two elements – a bar code proposed by IBM consisting of a series of parallel lines of varying thickness which represented the article number, and the printed human-readable number. Printing the number meant not only that it could be read by optical readers as an alternative but also that it could be key-entered if the bar code was unreadable. The details of both the bar code and the article number are contained in Part II, including how you can 'read' a bar code in the chapter 'Reading Between the Lines'.

Although the proposal from IBM was produced by Joe Woodland, the detailed work on the solution was done by others. The patent is in the names of George Laurer and David Savir, with credit given to Laurer for the development of the bar code and Savir for the check digit algorithm (which will be explained in Part II).

The decision was not welcomed by all. RCA had told the committee that, if its own symbol was not accepted, it would not develop equipment for any other symbol. After the decision, it was clear that there was no future for its own code and it withdrew from the market.

NCR continued to express severe reservations about the whole idea of symbol standardisation. But overall the retail industry and equipment suppliers increasingly accepted that a decision had been taken and moved towards the issues of practical implementation.

However, if one problem had been solved, another now had to be addressed. If the benefits of the new symbol and article number were to be achieved, the symbol had be printed on a substantial number of products and scanned in a substantial number of stores. To reach that position, manufacturers and retailers would be involved in significant initial costs. For retailers, they did not want to invest in scanning equipment until there were sufficient symbols printed on the products they sold. For manufacturers, they did not want to incur the cost of printing symbols on their products until there were sufficient scanners in stores to read them.
So who was prepared to risk an investment for no immediate return? The practical implementation of all this work and decision making looked as though it would take place over a protracted period. But then luck, coincidence or fate (the choice is yours) intervened.

The press announcement indicating the decision on the Universal Product Code and linked UPC Symbol was made at the same time as an independent announcement by the US Food and Drug Administration. Regulations were being introduced which would require nutritional labelling of food products. All grocery manufacturers would be required to re-design and re-label their products to meet these regulations. Faced by such a major change, many decided that the re-design of packaging might just as well incorporate this new UPC Symbol and Number, because it would be less expensive to do it at the same time than be faced by a further change to incorporate it at a later date.
And so it was that, during 1973, for the first time, supermarket customers in the US saw that some of the products in their shopping baskets had labels with these strange new lines and numbers.

The lines would revolutionise shopping not just in the US but also, as we shall now see, in Europe.

6. Longer and Longer

On 28th January 1974 I was happily celebrating my wife's birthday in Nottingham. I was completely unaware of a meeting which had just finished in Paris. Whilst I cannot remember what we did on that day, the decisions of the Paris meeting have been recorded. At a macro level, they led to today's world-wide bar coding and article numbering system. At a micro level, they would lead to my involvement with the subject for over 20 years and at a mini-micro level would lead to this book being written.

The Product Code Coordination Meeting was called by CIES (the International Committee of Food Chains) and chaired by its Director General, Frederic Treidell. It attracted 59 participants from 11 European countries and the United States, including directors from major retailers and manufacturers, trade associations, equipment manufacturers, consultants and the European Union. Its themes will now be well known to you, but Treidell's introduction added a distinctive European flavour. He expressed concern that in the seven EEC countries having different product code systems could prevent the free distribution of goods and so conflict with EEC legislation. Therefore it was essential that, at the least, maximum coordination was necessary to achieve compatible systems. In addition, if product codes were represented by a symbol, the symbol should have a common structure.

He then asked Tom Wilson of McKinsey to speak. Wilson had been involved throughout the developments in the US and had become a leading expert on the subject. He was asked to '*present an objective assessment of the current progress on coding and of the future direction*'. He referred to the decisions on article numbering and a bar code symbol which have been described in the last two chapters. He went on to talk about the progress which had been made in the US since the announcement of the UPC article number and bar code symbol some ten months earlier. By the end of 1974, it was expected that 50% of the units passing a checkout in a typical supermarket would have bar codes printed on them and by the end of 1975, this would be 75%.

(Both of these figures were weighted to reflect the number of units of each product sold, so the percentage of actual products coded was much lower.) He concluded that these high figures at such an early stage suggested not only that national numbering systems were achievable, for France and Germany had already shown that too, but also that a bar code symbol system was acceptable to the retail food industry.

In considering such potential development in Europe, Wilson referred to decisions taken by the Uniform Product Code Council in the US which would prove to be highly significant for the Europeans :

- In the course of developing a system in the US, it had not been and was not envisaged that the Universal Product Code would be extended beyond the United States and Canada.
- The UPCC has always been and still is ready to share technology, ideas and experience with Europe or any other part of the world.
- In addition, the bar code symbol belongs to the 'public domain' and its specifications are open to all.

Thus the message from Wilson was positive and optimistic, though he did add that much research was necessary in all the countries involved to examine costs, benefits and practical issues, a mouth watering prospect for McKinsey consultants across the continent.

Any such work would have to take into account two issues which did not need to be addressed in the US - agreement between different countries and agreement with the EEC. Dr Kempchen from the EEC started in encouraging fashion by stating that they did not want to get involved because this was a matter for retailers and their suppliers to sort out. But he then warned that this position could change. Different national coding systems could hamper the free flow of goods between countries, which was one of key freedoms established in the 1957 Rome Treaty.

He suggested that, as a minimum, it would be necessary to have some international system whereby each national coding system could be uniquely identified, possibly by a pre-code. And he warned that any national code should not compete for extension in territories outside its own borders.

There was a remarkable level of agreement in the discussion that followed on the way forward :

- All agreed that both food and non-food products must be included in any development (unlike in the US where the focus was initially only on grocery items). Many major European retailers sold both types of product and wanted a single system for all the items they sold.
- All agreed that a unique Symbol should be developed to represent article numbers (as in the US). Since symbols were not being used, there was a great opportunity and a great need to find a single solution at this early stage.
- But there was disagreement about the fundamental issue of the future of product codes. Whilst some thought that the ultimate goal of a single European system should be emphasised, others thought that the investment in existing national systems made that dream impossible and that maximum compatibility between different codes was the best that could be achieved.
- And so (as in the US) an Ad Hoc Committee was set up to progress these issues. The committee would consist of a retailer and a manufacturer from each of the 11 countries. In order to ensure that the decisions would be accepted and implemented, it was agreed that all members were to be company directors. Excluded from taking decisions, but available in an advisory role, were technical experts, trade association representatives and consultants.

To take this story forward, I must now introduce a figure who would dominate the bar coding world in Europe and beyond for the next 15 years. It is time to meet Albert Heijn.

Dr Albert Heijn (picture by Paul Huf)

*****

Many of today's major retail chains have spent most of their existence as family businesses, founded in the nineteenth or early twentieth century and developed during the past century by the founder and his heirs. During this period, many have changed to become public companies and may have lost any direct involvement by the founding family. In addition, many have extended their operations to other retail formats and other countries. But the names of these chains have remained familiar to new generations in their country of origin. Thus in the United Kingdom everybody will be familiar with Sainsbury's and Tesco, Marks and Spencers and Boots. And in the Netherlands everybody will recognise the name of Ahold.

Albert Heijn was one of the third generation co-owners of the Ahold family business, the largest supermarket chain in the Netherlands and one of the largest in Europe. He and his family were well known in the Netherlands. He was also well known in the retail industry, having spent much time elsewhere in Europe and the US seeking new ways of improving his Dutch business.

But he was also a hands-on retailer giving much time in visits to his stores. And it was the result of those visits that made him determined to find ways of reducing delays and resulting customer frustration at checkouts. This particular interest and his retail industry status made him an obvious choice as chairman of the Dutch Article Code Study Group.

He came to the Paris meeting in that role and left it as chairman of the Ad Hoc Council for European Product Coding. Given his influence in the next fifteen years, it is interesting to reflect on why he was chosen. Looking back on it thirty years later he told me it was unplanned, political and fortuitous. He received no warning of the suggestion either before or during the meeting, and the proposal at the end of the meeting came as a complete surprise. He thought Treidell had a political problem to resolve in finding a chairman. Given the strongly established but different positions of France and Germany, neither country would have been acceptable to the meeting as leading the search for compromise and agreement.

But Treidell was a Frenchman, based in Paris, influenced by the French retail industry and aware of their concern to protect their national system. So it was natural that he would seek a chairman who, at the least, was sympathetic to the French stance. And since Heijn not only had shown such appreciation in his comments during the meeting but also spoke French, he was a likely choice.

This analysis by him does not do justice to the fact that he had existing industry stature, knowledge and contacts throughout Europe and the US. Indeed others involved in these early years have said that he had already demonstrated that he had the interest, desire and time required to take an active international role. Such a person was required and such a person was found.

The first meeting of the Ad Hoc Council took place three months later in April 1974. The same 11 European countries were present – Austria, Belgium, Denmark, France, Germany, Italy, Netherlands, Norway, Sweden, Switzerland and the United Kingdom – with a total of 30 people, though Heijn himself was unable to attend. The morning session simply repeated the differences of view that existed and the chance of reaching some agreement looked remote. A lunchtime discussion of five of those present concluded that no immediate progress could be made.

So they scribbled down a set of nine questions which recognised the points of difference in a structured way and proposed to the afternoon meeting that a much smaller Working Party of six members should be given six months to suggest a way forward. At this point the acceptance of the proposal seems to have indicated a recognition of the impasse and vague hope that someone could resolve it rather than any optimistic expectation. But forces were at work which meant that within a year agreement was reached.

There was a flurry of activity. Immediately after the April meeting, the French and the Germans held a secret meeting in Cologne. They had the most advanced systems and had the most to lose by any decision on a new system. They had already had meetings to find some common solution, partly as a result of pressure from the EEC who had written stating that, '*The use of two different systems could waste some of the efforts which the Commission has made with regard to eliminating any kind of barriers to trade within the Community*'. Now, recognising the need to present a joint solution which protected their own systems, they devised a long and complex new product code structure that incorporated both of their codes.

However, the majority of those present at the first Working Party meeting on 20th June were not impressed when presented with a code which had so many digits.

They put forward an alternative much shorter code of eight digits (including a check digit), arguing that the total capacity of 10 million different product numbers was ample for the 11 or more European countries when the German and Austrian systems had only issued 160,000 numbers after many years.

The French Gencod Management Committee met two weeks later and their members were horrified at the way things were moving. Accepting any solution based on only eight digits would mean abandoning their eleven-digit code and investing in a completely new system.

They had spent two difficult years persuading companies in France to use the current eleven-digit code and could not contemplate the thought of going to them again with a further change. They decided that something must be done. So they sent Fernand Miot and Michel Laplane to the United States.

*****

There was only one difficulty with the decision to make this trip immediately – Fernand Miot's feet. A particular temporary problem meant that he would need to make the trip in carpet slippers. The effect of this on his American hosts is not recorded. Perhaps the success of the visit in July 1974 was due to the acceptance of a non-English-speaking Frenchman bringing with him either his country's eccentricity or their latest fashion craze. But perhaps it was due to the combination of Miot, the ex-diplomat director of Gencod and Laplane, the enthusiastic IT Director and technical expert from Nouvelles Galeries, a major French store chain selling both food and non-food products.

They were helped by the clearest of briefs. Their first objective was to demonstrate that the US experience of a long twelve-digit code and associated symbol had not presented any problems, and thus some European countries should not insist on a much shorter code. The second objective was to confirm that the US system would not be available to European countries.

Some of these countries had indicated that the use of the US system in Europe would be a good option. However, since the French code would need to change in those circumstances, they wanted to avoid that outcome at all costs. And the third objective was to demonstrate that the Gencod code could be translated into a bar code symbol similar to that in use in the US.

By the end of the third day of their four day tour they had achieved the first two objectives. Their meetings with key trade associations representing manufacturers and retailers had supported the comments made by Tom Wilson of McKinsey and John Hayes of Distribution Codes Inc, who was issuing supplier numbers, that the new code and symbol had achieved widespread acceptance. Furthermore, it had been stated once again that all efforts were being concentrated on US and Canadian industry, and that the Universal Product Code was definitely not going to be universal.

So on the final day, 17th July 1974, Laplane and Miot came to IBM in Raleigh, North Carolina to meet Joe Woodland. They left him with the problem of how to achieve their third objective. Could he suggest a numeric code that would incorporate the existing French and German codes, and that could be converted into a bar code symbol similar to the one developed for the US? Woodland indicated that he was confident not only that such a code could be found but also that Laurer would find a way of providing a linked bar code solution. Two weeks later Woodland and Laurer sent their recommendation to Paris.

*****

To assess Woodland's solution, consider the three coding systems in widespread use :

US - UPC	12 digits	0 ; Company (5) ; Product (5) ; Check (1)
France – Gencod	11 digits	Company (5) ; Product (6)
Germany – BanL	8 digits	Product Class (4) ; Product (3) ; Check (1)

Woodland's logic was simple. The code must be at least 11 digits long to incorporate the two European Codes. It could not be 12 digits long (and use the existing bar code symbol) because the US had allocated the extra first prefix digit for US systems. So why not make the code 13 digits to incorporate all three codes? He suggested the following could work :

US	13 digits	0	plus 12 digits UPC
France	13 digits	90 to 99	plus 11 digits Gencod
Germany	13 digits	xx000	plus 8 digit BanL
Other European code	13 digits	10 to 89	plus national 11 digit

This had several advantages. In the US, there would be no change because the new leading zero could be ignored and the 12 digit UPC would continue. In France and Germany, existing codes could be included in a longer European code. Elsewhere in Europe, it would offer an overall solution but give flexibility to each country too.

And the most important advantage of all was that Lauer had found a way of adapting the logic he had used in creating the US bar code representing 12 digits to produce a similar European bar code representing 13 digits.

It was everything the French wanted. Though of course there was one small problem. A month before, the Working Party had suggested an 8 digit European code. The French would now have to persuade both that group and Heijn's Steering Committee that a better solution was a somewhat longer code. What's more they would be proposing a length that had never been contemplated before – not 8 digits, not the 11 digits of Gencod, not even the 12 digits of UPC, but 13 digits. In many countries the number 13 is regarded as a sign of bad luck. So would this prove to be a case of unlucky 13?

7. Unlucky 13?

Albert Heijn soon realised that he had a problem. The second meeting of the Ad Hoc Council held in Paris in November 1974 was the first one which he was able to attend and take the chair. The presentation from Brian French, an Englishman who chaired the Working Group, produced a minute that the group had *'encountered a certain number of difficulties'*, a delicious understatement for which the writer deserves some acknowledgement. The reality was that the Group could not agree and were unable to make any recommendations. French pointed out that their role was to provide *'a technical approach to the problem in order to present a balanced point of view from technicians, fair and without bias.'* But, he complained, they were unable to do that because there was a minority non-technical view that the solution must incorporate their existing code without modification and be compatible with the UPC.

It is clear that the visit to the US by Laplane and Miot had reinforced the view of the French that the Gencod system must and could be protected. In other words they challenged the assumption that the outcome should be based on technical considerations alone. They argued that any pragmatic solution must be technically sound but should recognise practical and political realities.

The ball was being thrown back to the Ad Hoc Council. What did they want to do? It is clear from their comments that members of that meeting were as divided as those of the Working Group between an 8 digit solution and that suggested by the work in the US for the French of 13 digits. The minutes show the attitudes to a 13 digits solution including antagonism by the UK, scepticism by Switzerland, pragmatism by Italy and Sweden, and defiance by France. The following extracts from the minutes reveal the impasse faced by Heijn :

- *'It should not be forgotten that the more characters in the code, the higher the cost of computer and checkout storage.' (French – UK)*

- *'I would be interested to know how much it would cost Gencod or Ban-L to give up their systems and whether they are prepared to do so ; if not, how much will it cost to incorporate them into a European solution.' (Prout – UK)*

- *'The Working Group should look at this solution as well as other solutions and their cost. It could be that Gencod could decide to change their system for a European system.' (Husi - Switzerland)*

- *'I am in favour of the speediest possible effort on a European scale because existing systems are being introduced and no time should be lost.' (Orlandini – Italy)*

- *'It is impossible to estimate the cost implied in a European solution, but if Gencod do not give up their system, the solution of a 13 digit code will turn out to be necessary.' (Hoglund – Sweden)*

- *'The Gencod system is in a phase of evolution and implementation which is impossible to halt.' (Lieby – France)*

And as Heijn realised the problem he was facing in finding a solution that would command universal support, where did he stand in this debate? His only recorded comment to the meeting was that '*there was no doubt of the need for a code which would incorporate Ban-L and Gencod*'. He recognised the need for a pragmatic approach and was giving an indication of how the issue would be resolved some months later. However, this particular meeting reached no agreement and the Working Party was sent away to do further work on systems and their costs.

But before the Working Party reported back, Heijn was faced by another completely unforeseen problem – he was summoned to the European Commission in Brussels.

He had no idea why. In a letter of February 1975 he wrote, *'I have tried to find out what or who was behind the plans of the EEC authorities to call a meeting on the subject of article numbering'*. But his regular EEC contacts indicated that they were equally in the dark. He could not think of any reason why the EEC should be concerned at this time. Discussions were continuing on finding a European solution which would not result in any restraint on trade between EEC countries. Admittedly these talks had taken almost a year, but the EEC of all bodies should know that consensus is not always achieved quickly. So why did they want to see him?

When he got to Brussels he was confronted by an issue he had never even considered. They were worried about the United States! They had heard that a new system might be agreed which was based on the UPC number system and symbol in the US. Worse than that, the system had not only been developed in the US but was the work of a powerful, aggressive US company, IBM. Surely this would result in American companies having a major competitive advantage over European companies in this new market? Alarm bells were ringing in Brussels.

Heijn relaxed and reassured them. The development, decision and control of any new system would be in European hands. And, if the bar code symbol was finally chosen, the EEC should know that, although its original development had been done by IBM, the final agreed specification would be freely available to any equipment manufacturer who wished to develop products for printing and reading the symbol. Heijn had major problems to resolve but this was not one of them.

During the period leading up to the crunch meeting of March 1975, he was the subject of much lobbying. He was not happy with the progress made at earlier meetings and had even stormed out of one meeting at lunchtime in frustration to fly back to Amsterdam.

So in order to avoid another acrimonious and indecisive meeting, he wrote a note in advance to all members which he attached to the agenda. It said :

'In the past few weeks, a number of submissions have been made to me and to Mr French by members of the Ad Hoc Council, by representatives of the retail trade, and by other trade and regulatory bodies. These submissions have indicated that there is likely to be only one politically acceptable and feasible solution ...... that is Alternative 3 (the 13 digit code).'

The meeting accepted this proposal with varying degrees of enthusiasm. To appreciate the conclusion, it is best to ignore the official minutes of the meeting and to turn to Rolf Lindman, a Swedish member of the Council, who reflected on it in a letter to me 30 years later :

'When the French delegation suggested 13 digits would be needed, this was met with a lot of opposition. The French put it bluntly to us, 'If you don't accept this proposal we will not join the common efforts but go our own way with 13 digits because that is what is required for the future'. I don't remember how – but we finally accepted the proposal (or blackmail).'

That is not the bitter conclusion from someone bearing the scars of war, but rather a description of the reality of the outcome. For in the final sentence of his letter, Lindman says :
'For his foresight, all of us today have reason to be grateful to the French delegation and Michel Laplane.'

*****

We are nearing the end of this part of our story and the first section of this book. It may seem to some that I have laboured too long over these times and these decisions. But I make no apologies in doing so for three reasons.

Firstly, I think it is right to record the largely unheralded achievements of a group of individuals. Indeed relative brevity has meant that many other meetings, options, considerations, and names have gone unrecorded.

Secondly, I have wanted to emphasise that the current system which will be described in Part II did not emerge painlessly or inevitably. I invite you to consider the following. In 1975, Tom Wilson of McKinsey was asked to conduct a study of airline ticketing. Having been heavily involved with the retail industry achievements in article numbering and bar coding standards in the US and Europe, he was expected to be able to develop a similar industry-standard approach. After six months the project was abandoned because the different airlines were unable to reach a consensus. In a similar way, the banking industry had no common standards for many years and the move towards international standards has been laborious. Many other examples could be quoted. The past chapters have shown that agreement across groups with different interests, different starting points, different languages and different cultures before English had become a universal language of communication, was not inevitable. There was a real possibility that some other conclusion could have been reached or that no conclusion at all would emerge in both the US and Europe.

Finally, I note that, in this highly technical area which would require advances in computing, printing and scanning equipment, many of the ultimate decisions were taken primarily for political reasons. Who knows, there may have been a better mousetrap which could have been devised, but this mousetrap would be acceptable, would be widely used, and importantly would catch a lot of mice.

*****

As a teenager training to become a Methodist local preacher, I was often directed to the perceived wisdom of John Wesley's Forty Four Sermons and his general directions to his followers. In particular, I was reminded that, when preparing sermons, I should always say what I was going to say, then say it, and finally say what I had said. At a somewhat more advanced age, I am conscious not only that new things are harder to learn but also that old habits die hard. And so before moving on, I find it necessary (and I hope helpful) to summarise this part of the book.

From the late 1800s, the move from family-operated retail stores to larger shops employing increasing numbers of staff led to losses from 'sweethearting' and 'hands in the till' with a resultant search to find an alternative to the simple, but insecure, cash drawer. The development of the <u>Cash Register</u> was a major breakthrough in reducing business losses, leading to its widespread use by retailers.

However, by the late 1940s, even larger stores with even more customers and staff required a more advanced solution. The wider brief was to find a method of serving customers which would reduce the time taken at checkouts and thus reduce staff costs, whilst also further reducing losses from staff. The creation of the <u>Bull's Eye Code</u> offered an answer, though it was proving difficult to print consistently and the technical infrastructure to read and process the code was not available.

In the early 1970s, the retail industries in the US and in Europe separately decided that a common standard for product numbering with an associated machine-readable code symbol was necessary. It was anticipated that such a standard would bring a range of benefits to retailers, manufacturers and customers. Advances in laser reading and computer storage and processing technologies meant that a second major breakthrough could be achieved. The outcome was a <u>12-digit UPC number</u> and associated <u>Bar Code</u> in the US followed by a <u>13-digit number</u> and similar <u>Bar Code</u> in Europe.

Part I of this book has given the historical background to this achievement. But at this stage, although it has explained how article number systems and bar codes came to be introduced, it has neither explained all the details of that number system (what's in a number) nor explained the structure of the associated bar code (reading between the lines). If you are asking the question, *'Should I know by now what the bars and numbers mean?'* the answer is, *'No'.*

Part II will give you answers. It will give an introduction to the world of article numbers and show the meaning of the bars in a bar code. And it will explain the use of this system at the retail checkout as seen by customers today.

So for those of you who have travelled with me this far, and for those who have dived into the book at Part II, I will now answer the question, *'What's In A Number?'*

PART II – The New Retail World
(The Tip of the Iceberg)

8. What's In a Number

I have a choice of four different supermarket chains within ten miles of my home in Wirksworth, a small town in Derbyshire with a population of just over 5,000. At all of them, whether at a checkout or a counter or by self-scanning, the products I buy will be passed across a bar code reader. There is nothing unusual about that because it is now the norm in supermarkets around the world. As it happens, I rarely shop at any of them but buy most of my groceries at Kens and Dips, the two small family-run general food shops in the centre of the town. They use scanners to read the bar codes too. And if I visit any type of non-food store selling clothing, household goods, electrical items or DIY products the probability is that my purchases will be dealt with in a similar way. It feels as though this world of scanners, bar codes and article numbers is everywhere and furthermore that it has existed for ever. But, of course, it hasn't.

So when did it all start? Part I has given the general historical setting, but when did the first scanning store using today's bar codes actually open? It was an important enough event for the place and date to be formally recorded - Marsh's Supermarket in Troy, Ohio, USA on 26th June 1974. Furthermore, the first product to be scanned in the store on that day was also recorded – Wrigleys Juicy Fruit Chewing Gum (pack of 10). That particular product was not the first one to have a bar code printed on the packaging but just happened to be the first item to be taken from the basket of the first customer on that day. Its place in retail history is marked by its presence in the Smithsonian Museum in Washington.

Since that date, scanning has spread all over the world. After the United States, scanning came to Europe on 15th October 1977 in Germany and quickly extended to other European countries. Its introduction to other continents followed : Australia in 1978, Argentina in 1983, Japan in 1982, and Kenya in 1996. It is now seen in almost every country.

If you are interested in how old scanning is in your country and what the pioneering store was, a list has been produced specifically for this book and is shown in an appendix.

For such a common element of our lives, which is now taken for granted, it is interesting, although perhaps not surprising, that most customers of the system know little about it. This part of the book tries to increase that general knowledge by looking first at the printed article numbers, then at the associated bar codes and finally at what happens when the bar code is scanned at the checkout. But this is not a technical textbook, for such details can be found on many websites. So within the summary explanations of these subjects will be found some of the human side of their development. The starting point is the title of this book, 'What's in a number?'

*****

On the desk in front of me stands a jar of Cafedirect Instant Coffee. The labelling on the jar gives me as the consumer a lot of information. Some of this has been printed to meet legal requirements.

For example, I know that the weight is 100g and that the 'best before date' is April 2011. Other parts of the label have been included to help sell the product.

The statement 'Cafedirect won eight Great Taste awards in 2008' emphasises its quality. The Fairtrade mark indicates a particular ethical stance. But there is quite a large space given to something which serves no legal or marketing purpose but which is as important as anything else on the label. This is the 13-digit number and a set of 30 parallel lines of varying thickness above it.

The jar of coffee has the number 5 018190 009067. By chance I have another 100g jar of Cafedirect Instant Coffee in a cupboard – and it has the same number. But when and how did this particular product get this particular number? If you ignore the way the number is printed, which we will return to later, it has resulted from a process made up of four parts. By splitting up the number in a different way, this total process can be seen :

50 18190 00906 7

1. A global organisation called GS1 allocates to each of its member organisations (effectively country bodies) one or more two or three-digit GS1 Prefixes - in our example 50 is the United Kingdom.
2. The member organisation (country) adds to this a Company Number for each member company in its territory to produce a unique GS1 Company Prefix - in our example Cafedirect is 5018190.
3. The company allocates to each of its products its own internal item reference number to make a 12-digit number - Cafedirect gives Instant Coffee 100gm the number 00906 to make 501819000906.
4. Finally, the company calculates a check digit using a given algorithm, which will be explained shortly - for this product the check digit is 7.

The final 13-digit number 5001819009067 is unique to Cafedirect Instant Coffee 100g. It is unique not simply for any Cafedirect product, or even for any product in the United Kingdom. It is unique for any product sold anywhere in the world. As a result it is known within the retail industry as the Global Trade Item Number (GTIN).

To show this uniqueness I only have to look at the number printed on some other products :

Cafedirect Instant Coffee	100g	5018190009067
Cafedirect Instant Coffee	200g	5018190009289
Cafedirect Decaff Coffee	100g	5018190009197
Douwe Egberts Inst Coffee	200g	8711000055175
Teadirect	80 bags	5018190019172

The first three coffee products are all produced and sold by Cafedirect plc, so they have the same company prefix 5018190. But because they are either different sizes of the same product (100g and 200 g) or different products (coffee and decaffeinated coffee) they have different company assigned reference numbers 00906, 00928, 00919 and different check digits.

The fourth coffee product is produced by a different company based in a different country. In this case the Netherlands has been assigned the GS1 prefix 87 and has given the company producing the product (Douwe Egberts is part of Sara Lee) a company number. The internal reference number for Instant Coffee 200g and check digit have been added to produce a different and unique 13-digit GTIN for this product.

The fifth product, Teadirect 80 bags, is also produced by Cafedirect plc. Therefore, although it is a different type of product, it has the same company prefix 5018190 but a different company assigned number 01917 and different check digit.

Because I live in the UK, many of the products in my cupboards have numbers that start with the digits 50 – the GS1 prefix for the United Kingdom. But there are others which show that the number was allocated elsewhere. For example :

Swedish Glace Vanilla 750ml 73 13112060101   Sweden

Eswatini Lime Pickles   275g   60 05812000044   South Africa
Nobilo Sauvignon Blanc 75cl   94 14498400777   New Zealand

All of these products obtained their numbers by the same process. It is a remarkably simple system and yet it is able to ensure that any product in the world is identified uniquely. Of course, that assumes that the GS1 global organisation, the GS1 country members and individual companies do not give out any duplicate numbers for their parts of the overall number. But the beauty of the system is that two countries with different two or three-digit country prefixes can add the same digits for companies in their territories to create different GS1 Company Prefixes. Equally many different companies can allocate the same product reference number to their own products, but because it is only one part of the whole number, the full 13-digit Global Trade Item Number will always be unique.

Before we leave this introduction, we should complete the number by considering the check digit. Why is it necessary to have any check digit? Surely the 12-digit number is unique and should be sufficient. To understand the answer, take one of the 13-digit numbers above and type it out quickly 1000 times (or, if you prefer, think about typing it 1000 times). I suggest there is a strong probability that you would type it wrongly at least once and possibly several times. Although the article number (GTIN) is not often typed at a checkout, it does need to be typed many times during the life of a product. The numbering system could be reduced to chaos without some method of reducing the number of errors which result from mistyping.

What are the most common typing errors which you would make? I suggest that they are either that you will type one digit of the thirteen wrongly or that you will type two consecutive digits the wrong way round – for example 5018190009067 as either 5018190009077 or 5018190009076. The check digit system was devised using what is known as a modulo-10 algorithm to reduce the risk that typing errors would give an incorrect but valid number, possibly for another product. This is how it works.

Take the 12-digit Cafedirect product number 501819000906

Step 1   Add the alternate digits starting with the last, then multiply by 3
$6 + 9 + 0 + 9 + 8 + 0 = 32$; $32 \times 3 = \underline{96}$

Step 2   Add the remaining digits (working backwards again)
$0 + 0 + 0 + 1 + 1 + 5 = \underline{7}$

Step 3   Add the two answers from steps 1 and 2
$96 + 7 = \underline{103}$

Step 4   Find the smallest number to add to this to give a multiple of 10
$103 + \underline{7} = 110$ (a multiple of 10)

Result   7 is the Check Digit giving the GTIN of 5018190009067.

So whenever this number is used within a computer system, this calculation can be done with the first 12 digits of the number, and if the 13th digit is not the same as that which has been calculated, then the system knows that an error has occurred.

If you remain sceptical of these claims, I invite you to experiment by typing one digit incorrectly of the 12-digit number 501819000906 and then calculating the check digit for this number. I guarantee that you will not produce a check digit of 7.

However, in the second common type of error, typing two consecutive digits the wrong way round, the results are not quite as good, for in 10% of the 100 combinations possible, the check digit calculated would be the same and result in a valid 13-digit number. So the check digit system is not infallible but it does ensure that the vast majority of errors will be picked up immediately.

*****

This unique GTIN-13 number is most curious. During the 25 years I was involved with the subject, I lost count of the number of people who asked me what they could read in the number. We have seen that in one way the number tells you everything, because there is no other item which has the same number anywhere in the world. But in another way it tells you absolutely nothing. Consider some of the things that the GTIN might have been designed to tell you but does not tell you :

- Does the number tell you anything about where the product was manufactured or packed and what company did so? No. For example, all that the number for the Cafedirect product tells you is that the company selling it applied for its company prefix in the UK (because the first part of the number is 50). It does not tell you whether it was manufactured or packaged in the UK, and it does not tell you whether Cafedirect either has its own factory or has had it produced by another company working for it. As an example, a jar of Vegemite Yeast Extract in my kitchen has a number beginning 50 (applied for in the UK) but it says elsewhere on the label that it is 'Made in Australia'. There was a good reason for designing the number in this way, because otherwise any decision to change the country in which it was produced or packed would have required a new number and new labels.

- Does the number tell you anything about what type of product it is? No. From the outset, the international organisation decided that the number would be non-significant, in other words none of the digits could be used to indicate any grouping of products. This is unlike the German and Austrian Ban-L systems referred to in Part I, which had given significance to the number by rigidly categorising and managing every product. Again there were good reasons for this. Any such system would have required much greater central management and control, could have delayed the introduction of new products, and would have led to the requirement for a new number and label if any of the categories was changed in any way.

- Well surely the number indicates the price? No. Although the number and bar code are used to 'find' the price at the checkout, it is not included in the number. We will look at this in more detail later, but it is worth noting here the reason why the price was deliberately not included. Any particular product during its lifetime will almost certainly be sold at any one time with different prices in different shops, and the price will also vary within a shop at different times. If the price had been included in the number, the number and product label would need to have been different for each store chain selling at different prices, and changed every time there was a price change.

So in summary the GTIN is supremely unintelligent, it knows nothing and tells you nothing. But it is unique, allowing it to become the basis for a great deal of intelligence, as we shall see later. So what's in a number? The 'correct' answer is nothing – it is simply a non-significant number made up of four parts (numbering organisation, company, product and check digit) that allows an item to be identified uniquely. And that is how you should think of it in future. That may come as a bit of a blow. After all, both the book and this chapter are entitled 'What's in a Number' and the answer 'nothing' seems somewhat unsatisfactory.

But before you decide that this book is a misrepresentation and you should ask for your money back, please read on a little further. For the simple explanation of the GTIN-13 number above now requires to be expanded.

*****

I have listed some products found in my house together with the numbers printed on them :

Empress Pure Ceylon Tea	479 101641600 3
Artisan Black Olive Crackers	09 906800045 9
Twinings Echinacea & Raspberry	07 017711571 5
Mackays Rhubarb & Ginger Preserve	63 779300104 6
Baby Bio Plant Feed – 175 ml	50 37128 0
Schwartz Coriander Leaf – 6 g	50 92631 2

For when I look at the products found in my house, I find a somewhat more complicated picture than the one I have painted so far. Most of the products there do have the 13-digit number described earlier.

But a minority have different lengths of number. A few have a 12-digit number (such as the Artisan, Twinings and Mackays products) and rather more have an 8-digit number (like the Baby Bio and Schwartz products). However, whatever the number length, all of the products have a bar code printed above the number.

This particular mix of numbers is found in almost every country in the world. However, if by any chance you are reading this book in the United States or Canada, a different mix of numbers will apply.

There the majority of numbers will be 12-digit with some 8-digit and few if any 13-digits. The historical reasons for this were described in Part I of this book. The original system of numbering and associated bar coding was developed in the US and was based upon a 12-digit number (originally referred to as a Universal Product Code UPC-12 but now known as a Global Trade Item Number GTIN-12). The later European system, which was then extended outside Europe, was developed from the same ideas but was based upon a 13-digit number (originally referred to as a European Article Number EAN-13 but now known as GTIN-13). So whereas in almost all countries in the world the 13-digit number became the norm and is seen on almost all products, in the US and Canada the 12-digit number remains predominant.

But why are there any 8-digit numbers? The pioneers of our systems recognised from the outset that their decisions to have a relatively long standard 12-digit number (in North America) and a 13-digit number (in Europe) would present printing problems for many manufacturers and impossible requirements for some smaller products. So they created a parallel number system of only 8 digits which would allow these difficult products to have a shorter number and smaller bar code, occupying less space on the packaging.

However, if you have understood the four-part structure of the GTIN-13 number, then the explanation of GTIN-12 and GTIN-8 numbers will be straightforward. For the 12-digit number, the simplest way of dissecting it is to convert it into a 13-digit number by adding a leading zero.

So in the examples above, the numbers change from

Artisan Black Olive Crackers   09   906800045   9
                                   to 009   906800045   9
Twinings Echinacea & Raspberry   07   017711571   5
                                   to 007   017711571   5
Mackays Rhubarb & Ginger Preserve   63   779300104   6
                                   to 063   779300104   6

The process of understanding this number is identical to that we have seen before :

1.   The global organisation GS1 allocates each of its member organisations (countries) one or more two or three-digit prefixes – for GS1 US & GS1 Canada all the numbers 000 to 139 have been given, and in our examples 009, 007 and 063 are GS1 US.
2.   The country organisation adds a company number to make a GS1 Company Prefix, and the company adds its own item reference number to give the full number – for Twinings 007017711571
3.   The company calculates a check digit using the same algorithm as before – for this product the check digit is 5.

This slight complication is unfortunate, but is a result of history. However, it could have been much worse. For, whilst it may not be as satisfactory as having only one 13-digit number for everywhere in the world, it is infinitely better than what might now be the case – two completely different, unrelated systems and standards, with one for the US & Canada and another for the rest of the world.

Before we leave 12-digit numbers, the three products above may have suggested a further question. The companies concerned are all based in the United Kingdom. So why were they given numbers by GS1 US rather than GS1 UK, and thus have 12-digit rather than 13-digit numbers?

To answer that, it is necessary to understand more about the development of scanning systems and so the question will be left hanging until the end of Part II.

But moving on to 8-digit numbers, understanding the structure of the GTIN-13 and GTIN-12 numbers makes the GTIN-8 number very straightforward. The two products I gave as examples were

| Baby Bio Plant Feed | 175ml | 50 37128 0 |
| Schwartz Coriander Leaf | 6g | 50 92631 2 |

As before, the first two digits of the number represent the GS1 country organisation allocating numbers – 50 being the United Kingdom. And as before, the last digit is a check digit which is calculated using exactly the same formula as for the 12 and 13-digit numbers.

The remainder of the number (the middle section) is necessarily different. Because the overall number is so short, each two-digit country prefix only has a capacity of 100,000 short numbers (00000 to 99999) and each three-digit country prefix 10,000 short numbers (0000 to 9999) for all its companies and all the products they wish to number in this way. With this limitation, it is not possible to follow the two-stage process of the country giving the company a number followed by the company giving its own products a number. So instead these numbers are allocated singly by the GS1 country organisation to companies (unlike the longer numbers, where a company number is allocated and internal item numbers added by the company).

The same 8-digit number structure applies in all the GS1 countries, except for the US and Canada, where for historical reasons a different approach was taken. In order to reduce the complexity of this introduction to the subject, that particular structure is given in an appendix.

Preserving the short number capacity is very important. Therefore each country organisation applies a series of universal rules for circumstances in which a long number cannot be printed and a short number is allocated. In the early years some products were given short numbers at the request of companies where there was arguably sufficient space for the longer number. You may wish to examine any products you have with short numbers to judge whether this is the case today.

In doing so, you may find some examples of products with either short or long numbers which do not seem to fit precisely the descriptions I have given in this chapter. You may be thinking, 'But what about ....?'.

If so, I must confess that I have not told the whole story yet and there will be some more things to say about numbers and the GTIN a little later. With a bit of luck, all your questions will then be answered.

A fundamental understanding of GTINs has been covered in terms of 13, 12 and 8-digit numbers in this chapter. And even if the answer to the question, 'What's in a Number?' is 'Nothing', it is worth summarising the few key principles on which the system is based. For it is these principles that have been the reason why the system has lasted so long in a rapidly changing world and why it has been expanded to so many countries and industries for which it was not originally designed.

- Uniqueness – every product has its own worldwide-unique number
- Non significance – no part of the number has any particular meaning
- Persistence – every product keeps its unique number for its lifetime
- Decentralisation – the parts of the number are allocated at country and company level, consistent with global uniqueness

But this book was never planned to give a dry technical understanding of the retail world. So, on a somewhat lighter note, let me tell you a few stories about this strange numeric world.

9. What More Is In A Number

My interest in numbers goes back a long way. Arithmetic was my favourite subject when young and ultimately led me to a degree in Mathematics and even a period teaching the subject, though I do not believe I ever became a mathematician. But to those of you who can make such a claim, you will know that in number theory a lucky number is a natural number in a set which is generated by a sieve similar to the sieve of Erantosthenes that generates the primes. For non-mathematicians you may prefer, like me, an alternative understanding of what constitutes a lucky number given by the Mayor of Beijing.

He said before the Olympic Games in Beijing in 2008 that its success was absolutely guaranteed. His confidence had nothing whatsoever to do with the fact that it was being held in China, nothing to do with his belief in those who organised the event, and nothing to do with those who would take part. It would be successful, he said, because of the day on which it had been planned to begin – the eighth day of the eighth month of the year two thousand and eight. For eight is seen in Chinese culture by many as an auspicious number. And to ensure even greater success, the opening ceremony was planned to commence at 8 minutes and 8 seconds after 8 o'clock.

But what has this to do with our 13-digit number? Well, although I have said that the 13-digit number is non-significant, for some people there is no such thing as an insignificant number. Tan Jin Soon has been involved with bar coding as head of the national organisation in Singapore for many years and he explains :

'The Asian culture believes in the harmony of nature and the society. It also believes that luck plays a great part in the success of a person, a family and a company. In most Asian countries the number 8 is closely associated with luck and success. In 1987, EAN South Korea was the first country to request the particular country prefix 888. However, at the same time, the Singapore Article Number Council also requested the country flag 888.

In order to secure the country prefix 888 for Singapore, we thought of the idea '88 Olympic'. We then persuaded EAN South Korea to change their request for their prefix to 880, arguing that the whole world would look at a product from South Korea, see the number starting with 880, and remember that in 1988, the Olympic Games was staged in South Korea. So Singapore succeeded in getting 888 as its country prefix. Singapore is a very small island and needs the triple luck blessed by 888 to help us to be successful in implementing the system.'

The subject of lucky numbers not only influenced the <u>country</u> numbers allocated but also the <u>company</u> numbers given in some countries. In Malaysia, numbers are allocated to companies in sequence. However, the numbering organisation knows that if a company number contains the digit 4 (signifying death) then many companies will ask to be allocated a different number which does not have the digit 4. In Singapore, the numbering organisation has gone one step further. No company number containing the digit 4 is given out to anyone. And in addition no company number containing the lucky number 8 (signifying prosperity) is given out automatically – though these numbers can be requested and provided .... for an additional fee!

And lest you think that it is only Asian countries and companies that have brought some 'meaning' into part of the 13-digit non-significant number, consider the following tale of a company allocating the third part of the whole number, the internal company item reference number.

Many companies have their own product numbering systems which may or may not give some meaning internally to their products, and have used these same numbers within the longer 13-digit number. For new companies they may simply start with the number one for their first product and apply numbers sequentially thereafter.

But for one such company in France this was not enough. Therese Angue, who was head of the French bar coding organisation for many years, explained to the one-man company selling his first product that he could allocate any 5-digit number he wanted. As the owner wanted to use his Gallic flair and make the number personal, he eventually decided that the number would be the birth date of his child (fortunately able to fit into five digits). Therese was somewhat surprised to receive a call some months later from the anguished owner telling her, *'I have a problem. I have just introduced a second product and need to give it a number. But I only have one child!'*

*****

As the previous chapter and these stories show, the assignment of numbers is the composite responsibility of a global organisation, country organisations and companies. Since a particular product may be sold by many different retailers and even in many different countries, some global standards are clearly necessary to ensure each product is uniquely identified and duplicated numbers are avoided. This led the original Uniform Code Council (UCC) in North America and the equivalent organisation in Europe (set up as the European Article Number Association - EAN) to determine that they must apply a structure to a number, rules for when a product required a new number and further rules for when a number could be reused. (The two organisations UCC and EAN came together in 2002 and have used the name GS1 since 2005.) The standards produced by the earlier bodies have been largely unaltered, though a few changes have been necessary to accommodate the unexpected.

In 1975 when the number system was devised for the original 12 European countries, expansion to other European countries was anticipated. But nobody was thinking about such countries as South Korea, Singapore, Hong Kong and Malaysia and nobody remotely considered the possibility that numbers would ever be allocated in over 100 countries.

The group of 11 countries who had formed the Ad Hoc Committee, referred to in Part I, were joined by Finland and set up EAN. They were all from Western Europe. Just as UCC had set up an organisation to cover only the United States and Canada, so EAN saw their organisation as limited to Western Europe. And when they first came to allocate numbers to countries, this limited geographic scope was apparent. The 13-digit EAN number was structured initially in the following way :

Country (2 digits) ; Company and Item (10 digits) ; Check (1 digit)

Of the 100 two-digit Country Member prefixes, the numbers 00 to 29 were set aside for other purposes (which will be referred to later). This left 70 prefixes from 30 to 99 available for only 12 countries (and a few more European countries which might join later).

No country would want more than one country prefix, would they? After all, the 10 digits available for company and product gave a total of 10 thousand million products, more than enough for any country for many, many years. But if anyone did make such a request, there were plenty of numbers to go round. So in 1976 when France insisted that they must be allocated all of the eight prefixes from 30 to 37, nobody objected. The French case was very strong, for they already had a national numbering system of 11 digits which began with the numbers 0 to 7. They wanted to incorporate this within the new system. So if they added a 3 to the front of this and a check digit to the back, their existing 11-digit national number could automatically become a new 13-digit international number starting with a prefix for France of 30 to 37.

They were followed by Germany who wanted five prefixes from 40 to 44. The German case was less clear cut, for their argument was only based on the potential demand from manufacturers and retailers for company numbers. There was little to justify this request, and it is reasonable to ask whether such a demand would have been made or accepted if the French had only asked for one prefix.

But there was a spirit of great cooperation between the 12 countries and Albert Heijn was anxious to avoid any unnecessary confrontation Besides, there were still plenty of prefixes left, weren't there? Other countries were to follow. Italy were allowed the four prefixes from 80 to 83 and Austria the two prefixes from 90 to 91 without much consideration.

At this point a letter was received from a US organisation, Distribution Codes Inc, saying that they would want to take all the prefixes 60 to 69 out of the system (a demand that was later rejected). It caused Albert Heijn to consider what other demands on this two-digit prefix might emerge and to comment that *'although only 12 countries were at present involved, others might come in, even from outside Europe. Therefore it was important to have maximum number capacity'.*

This is the first formal reference that I have found suggesting that the European Article Number System might one day be considered for non-European countries. If this seems strange, it must be remembered that the world of the mid-1970s was very different to that of today. Although this international system was being developed, the bulk of retail trade at the time was based on products which did not even cross country boundaries. Within Europe, there was a recognition that, in the future, a somewhat higher proportion of trade would come from products exported between European countries. But nobody envisaged the global marketplace which exists today.

Albert Heijn's comments were partly triggered by a series of visits from a governmental group in Japan expressing interest in these European developments.

Whilst the original system had not been intended for extension outside Europe, it was recognised that if Japan became involved, one advantage would be that the use of bar code scanning in Japan would result in higher volumes of scanners being used worldwide with a consequent reduction in their price in Europe.
However, Albert Heijn's words about preserving number capacity had more significance than he realised at the time.

Whilst everybody involved expected some other countries to join the system, nobody expected that it would extend to over 100 such countries. But starting with Spain (in Europe) and Japan (outside Europe) the number of countries increased year after year. So how did the system and the organisation change to meet this new evolving geographical expansion?

Initially it led to an ingenious proposal from a Working Party in 1980. It said that it was '*inevitable and necessary that the system becomes worldwide*'. However, not only were there not enough two-digit prefixes for all these countries, there were not enough seats around the table at their meetings! Furthermore this relatively small group of people had developed relationships and ways of working, and some did not want the influx of a great number of countries to change the nature of these meetings. So they recommended limiting direct membership and '*inciting new countries to join one of the existing Coding Authorities*'.

The word '*incite*' was an appropriate one. The following year the New Zealand representative made the long journey to Europe. He suggested somewhat forcefully that the idea that New Zealand could not have its own organisation and prefix but should join the Australian organisation as a subsidiary was beyond belief! He won the day. And later that year it was agreed that new member organisations would be independent but, in order to preserve number capacity, they would be allocated a three-digit country prefix (giving 1,000 numbers) rather than a two-digit prefix (with only 100 numbers).

For the original countries it did not mean any change. For example, although the United Kingdom two-digit prefix of 50 did not change, it could also be thought of as the range of three-digit prefixes 500 to 509. But for new countries, the allocation of a three-digit number such as 888 to Singapore now allowed many more country prefixes to be reserved for future use. Of course, increasing the first part of the number from two digits to three meant reducing the second part of the number from ten digits to nine. But that still gave each country one thousand million possible numbers, albeit reduced from 10 thousand million.

And if they needed more, they could always ask for a second country number with a further capacity of one thousand million. So, in 1978, the system was formally changed to accommodate both 2-digit and 3-digit prefixes with the following number structures :

Country Prefix (2 digits) ; Company and Item (10 digits) ; Check (1 digit)

And

Country Prefix (3 digits) ; Company and Item (9 digits) ; Check (1 digit)

*****

Within this particular rigidity, the group of countries decided from the outset on a general principle of decentralisation. Clearly it was essential that some international standards were established and enforced, allowing trading across borders.
But where international decisions were not essential, it was agreed that national organisations would make their own decisions.

A fundamental example of this is seen in the structure of the Company and Item part of the 13-digit number above. In the UK, company numbers of 5 digits were allocated leaving 5 digits for the company's item reference number. In Finland, company numbers were 4 digits leaving 6 digits for the item reference. Each country decided on these elements and if they were different, what did it matter? As long as they each managed their own allocation process to companies efficiently, each 13-digit number would remain unique not just in their own country but worldwide. The only disadvantage to this flexibility has been to make the explanation of the number structure slightly more difficult for those writing about it, including the author!

As in so many things, the emergence of China as an industrialised society has impacted on this particular subject. In 1988 China joined EAN and was given the country number 690. Like all other countries at that time, it was able to manage its 9-digit Company and Product part of the 13-digit number according to its own requirements. It adopted the most common arrangement at the time of allowing four digits for companies, who could then give five digits for their item references. Such was the growth in the number of manufacturers in China, that within 18 months the 10,000 available four-digit company numbers were exhausted. So China was allocated an additional country number 691 in order to provide more company numbers, and has since been given 692 and 693 as well.

This explosive growth caused EAN to review once again the structure of the management of the 9-digit or 10-digit company and item part of its 13-digit article number. Pressure for more company numbers was being observed in many other countries, because other industries became aware that the use of these numbers, standards and technology could be applied and give business benefits outside the retail sector, for which the system had been originally designed.

In looking at the total number structure once again, it was quickly apparent that to give out company numbers such that each company could allocate up to 100,000 five-digit item numbers was unnecessary and that the global number capacity could be exhausted at some time in the future if this was not changed. So why did those responsible for the system allow this to happen in the first place? Since no company had 100,000 products and none even had 10,000, surely they could have had a number structure which gave fewer digits in the first place to the internal company item reference? The explanation is a simple one as we have already seen. Before EAN was formed, companies had their own item numbering systems, which could be as long as five or six digits. By allowing this structure, EAN had ensured that the vast majority of companies did not have to renumber their products but could incorporate their own internal number within the much longer EAN number.

And since nobody ever considered the extension of the system to so many countries, so many companies and so many other industry sectors, allowing five or even six digits for the item number did not present a problem.

But the wholly unexpected explosion in the use of the system meant that some further change to the structure was necessary. And, after much debate, it was decided to make this part of the number allocation even more flexible.

The two-digit and three-digit country numbers and final check digit would be unchanged. But the two parts of the central 9 or 10-digits (company and item) could be of varied length.

We have seen that any country already could be flexible in deciding the fixed lengths of company and item in their own country. But now they were given formal flexibility in these elements and could vary the lengths according to the companies applying to them for membership. For example, in a particular country some companies with many products or an existing numbering system could be given a five-digit company number allowing them to add up to 10,000 four-digit product numbers. Others might be given a seven-digit company number allowing for only up to 100 two-digit product numbers.

The concern expressed within the debate was whether this more flexible system would lead to errors and duplicate numbers, since each country would be responsible for controlling its variable-length company numbers to ensure that no duplicate numbers occurred. Was this change and slightly greater complexity justified? Absolutely – for the control system works. It is still remarkable that this simple system with billions of Global Trade Item Numbers read each year gives such a small number of errors.

*****

I said earlier that rules were also set for when a product required a new number and when a number could be re-used. The official specification of a requirement for a new number is *'whenever any of the pre-defined characteristics of an item are different in any way that is relevant to the trading process'.* If that summary is not particularly helpful, the alternative list of over 50 rules seeking to cover every eventuality is unlikely to prove 'a good read'.

So how can I simplify the rules for when a product requires to be given a new unique Global Trade Item Number? In practice, perhaps the best way of understanding this is to see each product through the eyes of the customer. So, any products which are of the same brand but with different sizes (such as Cafedirect Instant Coffee 100g and 200g) need different numbers. Similarly, you would expect that two tubs of a manufacturer's ice cream of the same size but with different flavours would need different numbers. And if a product is offered with, say, 50% extra free it becomes a different quantity item in the eyes of the customer and needs a different number. However minor packaging or ingredient changes do not require a new number. More will be said on this subject when we consider the scanning of a product at the checkout later in Part II and the delivery process from manufacturer to a store in Part III. At this point it is sufficient to note that there are global rules which should be applied, although the application of them is in the hands of each company numbering and labelling its products.

But can a unique GTIN ever be reused for a different product? Can a company ever reallocate the same item number as part of the total GTIN, specifically when the company knows that it is no longer producing and selling the product with the original number?

Back in 1977, concerns about running out of numbers were barely considered. However, it was recognised that there would be considerable wastage of numbers because, as products became obsolete, they were each buried with their unique global number.

This did not seem to make sense, particularly when every promotional, colour and size variant had its own number, but many of them only had a short life. Of course, the number should not be reused when a company ceased to supply it because stocks of it could be in stores for a long time after that. But it was agreed that there must be a system for reallocating numbers after a reasonable 'period of mourning'.

All were agreed that this period must be long enough to ensure that none of the product should still be available for sale in any store, anywhere. But how long was that period? Initial thoughts were rightly conservative and settled on six years from the date when a product was discontinued by a manufacturer. But this was quickly reduced to three years and, for many years, this proved to be a sufficient time period to allow duplicate numbers to be avoided and new number issue to be reduced.

Recently, the rules have been changed again. This has been caused by two factors, pulling in opposite directions. Firstly, as we shall see in Part IV, the GS1 system is now used in many other sectors of the economy than simply in retailing, where a product may take longer to disappear from view after its discontinuation, suggesting the time should be increased. But secondly, the vast use of numbers in the clothing industry, where products usually have a short life, would suggest the time should be decreased. So the new standards which recognise these points are 30 months for clothing and four years for other products.

In the last chapter, I answered the question, 'What's In A Number?' by saying that the correct answer is 'nothing'. It is just a unique number. In answer to the question, 'What More Is In A Number?' I am afraid the correct answer is still 'nothing'. And to make matters even worse, I have not told you the whole story and will come back to this subject soon. But at the end of these two chapters, I hope you will feel something of a 'dinner party expert' on article numbers. Even if you cannot be passed a bottle of wine and 'interpret the number' printed on it, you will have enough to say to cause a glazing of the eyes of your fellow diners and a recognition of the range of your knowledge.

By the end of the next two chapters, I hope you will be able to do the same on the subject of bar codes as we are about to 'read between the lines'.

10. Reading Between The Lines

This chapter is committed to the subject of how to read and understand a bar code. Not all attempts to deal with this subject have had this objective. And to introduce this explanation, I will refer to one alternative way in which bar codes were dealt with in the late 1980s by a comedy writer.

The television sketch produced a parody of the long-running BBC programme *Mastermind*. That programme consisted of four contestants each being given two minutes to answer questions on a subject of their choice followed by two minutes of general knowledge questions. The first contestant came forward, sat it the spotlighted leather chair and was asked for their chosen subject. He replied confidently, *'Bar Codes'*. The dialogue that followed started something like this :

Question Master   *You have two minutes to answer questions on the riveting subject of bar codes, starting now. How many lines are there in a typical bar code found in the United Kingdom?*

Contestant   *Thirty*

Question Master   *Correct. How many different bar codes are there in the world?*

Contestant   *How can anyone be expected to answer that? Nobody knows.*

Question Master   *Correct. What code is thin, thin, thin, thick, thin ....?*

Contestant   *A can of Heinz Baked Beans.*

Question Master   *Can you be more precise?*

Contestant	*A can of Heinz Baked Beans size 415 grammes.*
Question Master	*Thank you, that is correct. What is the code for a pack of 12 Nurofen Tablets?*
Contestant	*Thin, thin, thick, thin, thin, thin, thick, thin....*
Question Master	*No. Very close but the twenty-third line was thick not thin. An understandable mistake but you gave the code for a pack of 24 Nurofen Tablets. Hard luck.*

Well it isn't quite as straightforward as that, but I could have headed this chapter *Thick Thin Thick Thin* because that is how these lines are often seen on products and, to some extent, how they are read by scanners. As you look at a bar code it is important to keep in mind that it simply represents, in machine readable form, the number printed below it. For when these particular bar codes were developed, the twin requirements were to produce a number structure which would identify products uniquely and a machine-readable version of those numbers, which would allow them to be read quickly and accurately.

So both are seen on products and, although it is the bar code which is normally read by a scanner, the number itself is always available as a backup which can be entered manually if the code is printed or attached badly.

We will take the two bar codes which are the symbols corresponding to the 12-digit and 13-digit Global Trade Item Numbers. Whilst reading two codes will make this chapter a little longer and somewhat more complicated, it is necessary for two reasons.

Firstly these are the codes which are on the majority of products. But secondly, the coding and reading system which was developed by Joe Woodland and George Laurer for the 12-digit UPC code (first known as UPC-A and now GTIN-12) in North America had to be modified following the visit of Michel Laplane and Fernand Miot to create a 13-digit EAN code (first known as EAN-13 and now GTIN-13). To understand this longer code, we must follow a path via the shorter one. We will do this in four stages :

- Finding the Code on the Packaging
- Reading the Right-Hand Side of the Code
- Reading the Left-Hand Side of the Code for the 12-digit UPC
- Reading the Left-Hand Side of the Code for the 13-digit EAN

*****

5 012345 678900 >

6 14141 00086 9

< 5012 3452 >

0 012345 7

Finding the Code on the Packaging - Guard Bars

Both codes have 30 lines or bars, which should be seen as 15 pairs of bars. Both have three pairs of longer bars - at the start of the code, in the middle and at the end. In the 13-digit code these are the only long bars. In the 12-digit code there are two other pairs of long bars, but these have been made longer for design purposes only and have no 'reading' significance. The three identical pairs of 'thin' longer bars are referred to as guard bars and mark the beginning, middle and end of the code. The purpose of these guard bars is so that a bar code scanner searching the packaging of a product can know that, when it 'finds' these distinctive bars, it has found the bar code and not some other printing on the pack.

Back in 1974, there was great concern that the pressure on designers to produce attractive labels with legal and marketing information could mean that the printing of the bar code and number would be squeezed into a small space. As a result, the scanner would have difficulty in finding the code, muddling the code with other printing and leading to a lot of problems at supermarket checkouts. So, it was decided that a minimum space was needed before and after the code to ensure that it could be found and read. This space was termed the *Quiet Zone* in which no other printing should appear. But would everyone keep to any rules that provided this minimum of space? So the code designers for UPC printed the first and last digits of the 12-digit number outside the lines to make sure that this space was allowed. The remaining 10 digits were printed in two groups of 5 digits underneath the bar code.

The same code designers had a bit more of a problem with their 13-digit EAN code. So they decided to put the first digit outside the front of the bars and 6 digits underneath each half of the code, but simply define the space width to be left after the code in order to prevent scanning errors. Some years later it became obvious that the rules defining this minimum space were not always followed, and reading problems were occurring as a result. Therefore, you will often, though not always, see a small arrow printed after the code and 13-digit number on many products, the purpose of which is to make sure that sufficient space is left at the end of the code.

Having found the code with the help of the guard bars and the space around the code, the scanner can proceed to read the code. It only attempts to read the bar code and will <u>not</u> read the number. But the guard bars not only help a scanner to find the code. They have a second critical function. The thickness of each of these lines and the space between them gives the scanner a 'standard width' which it will use when it reads the variable thickness of the other bars.

Between the outer guard bars and the centre guard bars, in each half of the code, are 12 bars which will be read as 6 pairs of bars.

For the UPC Code this will be relatively straightforward, for each of the 12 pairs of bars represents one of the 12 digits of the number. For the EAN Code it will be more complicated, because there are 12 pairs of bars which must represent a 13-digit number. So a little ingenuity will be necessary, as we shall see. But before that, we can start with common ground for both codes.

*****

Reading the Right-Hand Side of the Code

In Part I, Joe Woodland's Bull's Eye Code found a way of reading a bar code from any direction. But a different solution was necessary to ensure that the UPC and EAN sets of straight lines could be read accurately, however they were presented to the scanner. To achieve this omni-directional reading, a system was devised which treated the two halves of the codes in different ways, using different formulae. By doing this, the scanner could read each half of the code, check against the formulae and know which half was the left and which was the right. For the right-hand side of the code, the same formulae are used for both the UPC and EAN codes, whereas for the left-hand side of the code, they vary. So we start with the (relatively) easy side, the right-hand side.

The six pairs of bars in the right-hand half of the code represent the last six digits of both the UPC and EAN numbers, and each digit from 0 to 9 is represented by a unique pair of bars. Although in the *Mastermind* sketch the bars were referred to as *thick and thin,* when you look more closely at the bars you can see that they vary in thickness.

Value of Character	Number Set A (Odd)	Number Set B (Even)	Number Set C (Even)
0			
1			
2			
3			
4			
5			
6			
7			
8			
9			

Looking at the table for Set C, you will note that each number is formed by seven printing spaces, each of which can either have ink (shaded) or no ink (unshaded). Although the table shows these as wide spaces, the actual printed bar code is of course much narrower. For any number the result is two vertical black lines which can each be of one, two or three printing space widths.

For example the number 0 in Set C is represented by a line of ink width three, followed by a space of width two, a line of ink width one and a space of width one. So whenever the number 0 is shown in the right-hand side of the bar code it will always appear as a thick line (of width three) and a thin line (of width one). This means, for example, that if the six digits on the right-hand side were 000000, there would be six identical pairs of bars. Similarly, looking at the table, you will realise that if the six digits on the right- hand side were 123456, there would be six different pairs of bars.

Although all the digits 0 to 9 have a different representation, you may have noticed in Set C that there are two things which are common to them. Firstly, all of them start with a printing space with ink and end with a printing space without ink. If you think about this for a moment, this means that a scanner always knows it has 'found' the start of the first bar of a digit (a 'with ink' first printing position) and that a space is always left after the end of the second bar of a digit, (a 'no ink' seventh printing position) which separates it from the next pair of bars. Secondly, all of them have a total of either two or four ink spaces. This is referred to as 'even parity' and its importance will become apparent when we look at reading the left-hand side of the code.

Going back to the *Mastermind* sketch, the lines in a Bar Code do appear as Thick and Thin. So we could say that the lines could be Thin (1 space with ink), Medium (2 adjacent spaces with ink), Thick (3 adjacent spaces) and Very Thick (4 adjacent spaces). Then the digit 0, which is made up of 3 Ink, 2 No Ink, 1 Ink and 1 No Ink, would look like a Thick bar (3 spaces) followed by a Thin bar (1 space).

Using I - Ink and N - No Ink, the digits 0 to 9 for the right-hand side of the code could be thought of in this alternative way and provide an alternative way of representing Number Set C :

	Right Side	Number Set C	
0	I I I N N I N	Thick	Thin
1	I I N N I I N	Medium	Medium
2	I I N I I N N	Medium	Medium
3	I N N N N I N	Thin	Thin
4	I N I I I N N	Thin	Thick
5	I N N I I I N	Thin	Thick
6	I N I N N N N	Thin	Thin
7	I N N N I N N	Thin	Thin
8	I N N I N N N	Thin	Thin
9	I I I N I N N	Thick	Thin

Unfortunately, this way of reading means that some of the pairs of bars are the same, for example, 'Medium Medium' for numbers 1 and 2. But as you will see in Number Set C, the precise position of these two 'medium' bars within the seven printing spaces is different, making 1 and 2 unique to the scanner.
I am afraid that this means it will be more difficult for you to distinguish between the four digits 3,6,7,8 which are all 'Thin Thin' with the naked eye, but you may be reassured to know that the scanner does not have the same problems as you.

For some of you, I may have laboured this explanation. For others, I hope it throws some light rather than dust on the subject. But, as was noted earlier, in one sense the bar code symbology is not important. It is the number that is critical, whereas the bar code simply gives a speedy and efficient way of reading that number. However, you may be one of a small minority who is sufficiently interested in this subject to learn these number representations. Some 25 years ago, when I first became involved in this subject, I took a decision that I would not bother to do so. It is a decision I have never regretted.

So the six digits on the right-hand side of the UPC-A and EAN-13 codes have all been represented by pairs of bars according to the table for the digits 0 to 9 above. It <u>always</u> works.

However, if you try to use this table for the left-hand side of any code, you will find that there is <u>never</u> any match between the printed number and the pair of lines. As we shall see, the digits 0 to 9 are given a <u>different</u> pair of bars when they appear on the left-hand side of the code. This has been done deliberately so that the scanner knows when it 'translates' the six pairs of bars which are found in Number Set C, it must be reading the right-hand side of the code.

*****

Reading the Left-Hand Side of the Code for the 12-digit UPC

The left-hand side of the UPC symbol will be much easier to understand. Here the digits 0 to 9 are represented by simply taking Number Set C for the right-hand side and reversing all the dark and light printing spaces to give Number Set A. In other words the patterns of <u>dark</u> printing spaces for digits on the right-hand side become the pattern for <u>light</u> printing spaces on the left-hand side.

Using the alternative way of thinking about the thickness of the lines gives the following :

	Right-Side (C)	Left-Side (A)	Left Side Lines (A)	
0	I I N N N I N	N N N I I N I	Medium	Thin
1	I I N N I I N	N N I I N N I	Medium	Thin
2	I I N I I N N	N N I N N I I	Thin	Medium
3	I N N N N I N	N I I I I N I	Very Thick	Thin
4	I N I I I N N	N I N N N I I	Thin	Medium
5	I N N I I I N	N I I N N N I	Medium	Thin
6	I N I N N N N	N I N I I I I	Thin	Very Thick
7	I N N N I I N	N I I I N I I	Thick	Medium
8	I N N I N N N	N I I N I I I	Medium	Thick
9	I I I N I N N	N N N I N I I	Thin	Medium

111

Several things are worth noting. All of the digits are represented differently according to whether they are on the left or right-hand sides. All of those on the left start with a light space and end with a dark space (the reverse of what we saw for the right-hand side). And all of those on the left have three or five of the seven printing spaces as dark spaces (referred to as odd parity) against two or four on the right-hand side (even parity). The critical importance of this last point is that the scanner knows which half of the bar code is the left-hand side, because all of the digits on the left will have 'odd parity' and all on the right 'even parity'.

And that allows me to offer you good news and bad news. The good news is that, if you have followed the explanations above, you are now able to use either of the tables above to read a 12-digit bar code. The bad news is that there is an additional complexity to be understood before you can read the 13-digit EAN code.

*****

Reading The Left-Hand Side of the Code for the 13-digit EAN

How could the logic of the bar code designed specifically for a 12-digit article number be modified for a bar code of a 13-digit number when there were only 12 pairs of bars?
To change the fundamental structure of the bar code from 12 pairs of bars to 13 pairs of bars was too complex. But was there some ingenious way in which 12 pairs of bars could represent 12 numbers and allow the thirteenth number to be 'calculated'? It was not good enough to simply represent the first 12 digits of the 13-digit number and let the scanner 'calculate' the check digit. The check digit had been built into the number to 'find' invalid numbers caused not only by typing errors but also by printing or scanning errors. So could a system be devised for representing the last 12 digits of the 13-digit number and let the scanner 'calculate' the first digit in a way that detected printing errors?

If all of that sounds simple and logical, it is a conclusion that I do not think I could ever have reached.

112

And if you have doubts about the ingenuity of the solution which George Laurer devised, you will have them dispelled by the details of that solution which follow.

To read 12 digits and calculate the thirteenth was achieved by a three stage process. Firstly, the right-hand side of the code would follow the same pattern for 12 and 13-digit numbers, as we have seen above. Secondly the digits 0 to 9 on the left-hand side would be represented not in just one unique way (as in Number Set A for the UPC code) but in either of two ways (Number Set A and Number Set B below). Thirdly, the particular combination of how these Number Sets A and B would be used for each digit would determine what the unrepresented first digit must be.

If you look back at this table you will see that a very simple change was made. Whereas Set A had been created by reversing the shading of each element (dark and light) in Set C, Set B was created by reversing the direction of Set C.

The two possible representations are shown below.

	Left Side (A)	Left Side (B)	Left Hand Lines (B)	
0	N N N I I N I	N I N N I I I	Thin	Thick
1	N N I I N N I	N I I N N I I	Medium	Medium
2	N N I N N I I	N N I I N I I	Medium	Medium
3	N I I I I N I	N I N N N N I	Thin	Thin
4	N I N N N I I	N N I I I N I	Thick	Thin
5	N I I N N N I	N I I I N N I	Thick	Thin
6	N I N I I I I	N N N N I N I	Thin	Thin
7	N I I I N I I	N N I N N N I	Thin	Thin
8	N I I N I I I	N N N I N N I	Thin	Thin
9	N N N I N I I	N N I N I I I	Thin	Thick

Deciding which of these tables to use for each digit depends upon the unrepresented first digit of the 13-digit number as follows :

First Digit	6 digits in order use Table
0	A A A A A A
1	A A B A B B
2	A A B B A B
3	A A B B B A
4	A B A A B B
5	A B B A A B
6	A B B B A A
7	A B A B A B
8	A B A B B A
9	A B B A B A

Two things are worth noting in comparing Tables A and B. Both have all digits starting with a light space and ending with a dark space. But whereas Table A digits are all Odd Parity (3 or 5 dark spaces), Table B digits are Even Parity (2 or 4 dark spaces). So when a scanner reads a UPC code, all the left hand side digits are odd parity and all the right hand side are even parity. But for an EAN code, the left hand side are a mixture of odd and even parity and all the right side are even parity.

This difference allows the scanner to determine whether it is reading a 12-digit UPC code or a 13-digit EAN code and to use the appropriate logic to deduce the number.

And that is it. To say the least, these rules do not make it easy for you to translate a bar code into a number. Fortunately, the software behind the scanner does not have your difficulties. And the rules do not make it easy for me to translate this system into a readable chapter. So I hope the summary below is helpful before we move on to easier and lighter subjects.

* * * * *

Reading The Code - The Scanner

1. Finding The Code - The scanner 'reads' the packaging and knows that it has found the code when it discovers a complete set of 30 bars with pairs of 'guard bars' at the start, middle and end of the code.
2. Finding The Width of a Printing Space - The guard bars are all of one printing width with one space in between them. The scanner will use this width to determine the relative width of all the other bars. It does not affect this calculation if the bar code is read at an angle.
3. Finding The Parities - By finding which half of the code has all pairs of bars as even parity and therefore the right hand side, the scanner knows whether it is reading the code the correct way up or upside down.
4. Translating the Code - The scanner translates the pairs of bars into 12 digits using Tables A, B and C and determines whether it is a UPC or EAN code.
5. Deriving the First Digit - If the left-hand side has mixed parity, the particular combination of odd and even parity for the 6 pairs of bars on the left-hand side allows the first digit to be calculated to make a GTIN-13 number.
   For the UPC code, since all these pairs are of odd parity, it has already read the GTIN-12 number.

At the end of an earlier chapter I hoped that you had become an article number 'dinner party expert'. To that qualification, I hope is added that of 'bar code reader'. And if 'reading between the lines' has proved as difficult as I feared, perhaps 'reading more between the lines' will offer some light relief.

11. Reading More Between The Lines

*'Let him that hath understanding count the number of the beast :
for it is the number of a man ; and his number is six hundred, three
score and six'.*

You may or may not know that this quotation appears in the King
James translation of the Bible. But you may wonder what possible
reason there can be for quoting it here and what possible link there
can be to 'reading more between the lines'. Those involved in the
early days of article numbering and bar coding wondered too.
They had plenty of problems to solve. They had plenty of
criticisms to address from those who did not want the changes
brought about by these new technologies. But they were aghast
when their new bar codes came under attack from a group of
biblical fundamentalists. For, it was claimed, these new bar codes
displayed the number 666 and this 'mark of the beast' signified
anything from the evil of the new system to a sign of the coming
end of the world.

As you look at the bar codes on products, you will be somewhat
surprised at this conclusion for you will not see any indication of
the number 666. However, your patience in completing the last
chapter is about to be rewarded. For, as a newly qualified expert in
bar codes, you will now be able not only to understand why these
claims were made but also to refute them.

So where is the number 666? The claim is that the the three pairs
of guard bars at the beginning, middle and end of the bar code each
represent the number 6 and that taken together they represent the
number 666. But why would each of these pair of lines represent
the number 6? Because, the claimants note, when the number 6
appears on the right hand half of a bar code it has a pair of lines
(Thin Thin) that are identical to the guard bars. And surely that is
proof enough?

But some claimants took the matter further. They examined these bar codes for any further evidence that the 'mark of the beast' was clear for all to see. These 'special studies' noted that it was highly significant that each half of the bar code contained 6 pairs of bars, and that the central pair of guard bars was unnecessary but meant that instead of two sets of guard bars representing 66, three sets led to 666. Finally, if any further proof was necessary, why had the EAN 13-digit code been designed to print the first digit in front of the bars? Obviously so that 6 numbers would appear under each half of the code.

As we have seen, the guide bars serve particular purposes and do not represent any number. All three pairs consist of two thin lines separated by a single printing space. And it is true that the number 6 on the right side of the code is the only digit represented by two thin lines separated by a single space (see Number Set C). However the number 6 has a different representation on the left hand side for UPC (Set A), though the other left hand digit 6 for the EAN Code is also two thin lines separated by a single space (Set A). Finally we have seen how the UPC Code came to be 12 digits, why it was split into two parts to read the code, and why the design of the EAN Code put the first digit in front of the bars to ensure that space was left around the code.

For those who designed and introduced the system, this description of their work seemed extraordinary. So they explained again and again the background to the bar code and the reasons for its detailed design. Although the subject is raised less frequently than in the past, there are still many internet articles repeating this story. And no doubt those who have seen, and still see, a conscious or unconscious hand at work in this design will continue to do so.

But this is not the only way in which bar codes have acquired uses for which the founders were unprepared.

*****

Though perhaps they should have anticipated that the art world would welcome their introduction. After all, modern art always makes use of a whole range of contemporary objects and images. So, as bar codes began to appear on a wider and wider range of products, they could be used as an example of the modern world of technology and consumerism, or simply as an artistic device.

One recent example of this is in the work of Scott Blake who has established a business showing portraits and other images from associated bar codes. A picture of Madonna has been made up from the bar codes appearing on 107 of her albums. As such it can be viewed in different ways as an illustration of today's commercial world.

Sylvia Grace Borda uses Andy Warhol's pictures which refer to Campbell's Soup labels as a base but uses bar codes rather than the label to create objects which are '*distant and estranged*'. She says, '*The reception of the barcode as an artform parallels Popart's reception – it is seen as cold and mechanical*'.

These somewhat negative images of bar codes were graphically illustrated by Mad magazine who devoted the whole of the cover of issue No. 194 to the subject. Half of the cover contained an enlarged bar code. Above it was the headline, '*Hope this issue jams every computer in the world ....*' and below it the words, '*...for forcing people to deface products with this yecchy price code symbol.*' Making the front cover of a magazine was not limited to Mad. In January 1985, Time Out magazine showed a multi-coloured bar code set without a picture frame and the caption 'What Price Art?'

(source : author)

But not everyone was as serious and codes can be seen as the basis of themes in many cartoons and on postcards. These often show a person, an animal or an item depicted with a bar code. The first one that I can recall being sent showed two sheep on a Welsh hillside talking to each other beside a wall, with a listening but confused sheep on the other side of the wall. The caption read, 'To baffle eavesdroppers, Wendy and Molly sometimes conversed in baa-code'.

119

This publicity of bar codes was a reflection on how quickly they had become an established fact of everyday life. Their depiction in so many places was a source of satisfaction to those who had been part of this development. However, if making the front pages of magazines felt like something of an award, making the front pages of national newspapers was quite different, as we shall see shortly.

*****

The impact of the growing appearance of bar codes was felt by those managing the system in another way. The first example of the phenomenon was in France. The children of a school and their parents started to collect the bar codes from products. The more they could collect, the more money would be given to the school. Now this is not unique because supermarket chains have often linked purchases in their stores and the collection of vouchers with providing equipment for schools.

Unfortunately, in this case, neither the supermarkets nor the French EAN organisation knew anything about it. By the time that they became aware of what was happening, a considerable quantity of bar codes had been accumulated and the rumour had spread to other schools and other areas. For EAN France this was a painful diversion at a time when it was fully occupied with both resolving the many problems associated with the introduction of the new system and managing its expansion to more retailers and more manufacturers. Whilst it was encouraging to know that the bar code had become an accepted part of retailing, it was disturbing for them to realise that their announcements that no such scheme existed and no such money was available would not be particularly well received by those whose garages contained boxes of bar code labels.

This particular phenomenon was repeated in a few other countries. But the problem which was seen in almost every country was that of the 'rogue company'.

Perhaps it was inevitable when so many companies were being told by retailers to add bar codes to packaging, from the very large multinationals to the sole traders with only one product, some would do this in non-standard ways. Whilst the readability of codes should have been assured by the use of specialist companies who produced *film masters* with the number and bar code for use in the printing process, some bypassed this system. And so stories of bars which were hand-drawn, copied from existing products, and produced in colours which could not be read can be found in almost every country. Fortunately the number of these instances was relatively small, but they were often not discovered until a code would not scan.

So at this stage it is worth pausing to consider how the system is managed to ensure that problems of this sort are not more widespread.

*****

And the starting point is that readability is more or less taken for granted. As the supermarket trolley is emptied, as the conveyor jerks forward, and as the products are scanned, it is assumed that this process will be efficient. Of course there are the occasions when the bar code will not scan first time, and fewer instances when it will not scan at all, but overall we do not expect this to happen. It was not always like this and it has not happened automatically.

From the outset, the US and European organisations knew they had this high hurdle to jump - minimising non-reads was necessary to reduce staff time and customer frustration at checkouts. But bar codes would be printed initially on millions of products, by thousands of companies, and on many types of material. They would be read by many retail chains using a variety of scanning devices. How could you ensure a high overall quality in that environment?

Perhaps the timing of the introduction of the systems helped. There was at that time an understanding brought from Japanese companies of Quality Management. So it was recognised that poor quality at the start of a process led to exponential cost increases during the later stages of the process. In this case, a bar code printed wrongly such that some or all scanners could not read it, could result in many thousands of non-reads throughout the life of the batches of the product which contained that particular bar code label.

Enormous efforts went into specifying precisely how the bar code should be printed and controlling its quality. To give you some idea of the work that went into achieving high-quality fundamental design standards, I quote Neville Hughes (one of many specialists who spent much of their working life passionately committed to improving standards) who worked for the Metal Box Company :

'Specific printing applications required detailed studies. The direct printing of three-dimensional containers is a good example. Cans as used for beer, Coca Cola and other beverages are printed in-line with their manufacture and speeds – even in the 80's - of over 1,000 cans a minute. To compound the problem, many designs used the metallic base and hence white spaces as well as coloured bars had to be printed. Similarly, plastic containers posed a few challenges including tapered transparent ones and coloured bottles.

Manufacturers wanted symbols as small as possible but retailers wanted maximum omni-directional scanning. Metal Box prepared a wide range of truncated samples and Fine Fare closed one of their stores on a Saturday afternoon. My 10 year old son spent two hours feeding a scanner! The guidelines were strengthened.'

It is interesting that Neville Hughes uses the word *guidelines* rather than *rules*. For, as in the management of many of their standards, UCC and EAN operated by consensus, imposing standards where necessary but leaving flexibility elsewhere in the expectation that any problems would be resolved between trading partners. This can be seen in a few examples.

Where should the bar code be printed on the packaging? You must have been at a supermarket checkout where the operator has searched the product to find the code.

So surely a simple inflexible rule, such as always printing on the top right-hand corner of a label, would have eliminated these problems? Such a discussion did take place. It was soon pointed out that in some countries and some retail stores, counter-top flat-bed scanners would be used such that a code on the bottom of each product would be the most efficient. But others would either have scanners which were vertical and read codes from the side, or used a hand-held scanner. And there were so many different types and sizes of products with so many print surfaces, printing processes and limitations on space, that any overall rule would never have been acceptable to manufacturers. So you will find bar codes in many different positions on products today. Whilst that might seem inefficient, in practice the improvements in scanning technology have meant that the checkout process has not been adversely affected by this decision.

Some other early decisions regarding printing can be found on today's products. The bar code is always printed as dark lines against a lighter background, never the reverse. And though most bar codes are printed as black lines against a white background, artwork designers do have the freedom to print using a wide range of other colours. Whilst the specification of the bar code must ensure that it can be read by a scanner, the same is not true for the printed number. From the outset, it was assumed that the bar code was the machine-readable version of the number : the number itself would not be scanned but could be entered into computer systems manually if necessary. Most numbers are actually printed with a particular font known as OCR-B, but this is not mandatory and other fonts can be seen.

A further interesting aspect of printing concerns products which are sold in packages that are cylindrical (eg cans and tubes). If the bar code is printed round the can, distances between the lines are distorted when it is scanned and the code may not be read correctly.

Therefore the specification for printing indicates that it is preferable but not obligatory for the bar code to be along the can, but with a detailed rule set out for small tubes to ensure a it can be scanned correctly.

*****

In the early years of scanning, many retailers faced three particular problems. Firstly, a significant number of products were not 'source-marked'. They did not have a bar code printed on them, and thus the store had to separate out such products when they were delivered and print a unique code and number on labels to be attached to each item. This solved a problem at the checkout by ensuring that every product presented had a bar code. But it created potential problems in the back of the store where staff had to ensure that the correct number was always found for the product and the correct bar code was always printed. Fortunately, the steady growth of source-marking reduced this work and the number of resulting errors.

Secondly, some products had bar codes which could be read by the manufacturer's scanner but did not scan on one or more retailers' different scanners. These were often detected before they reached the shop floor by providing a spare scanner in the back of the store to check readability. But that also created a problem. Who was at fault and who was responsible for sorting out the printing issue? In some cases, a retailer could show that a manufacturer had not printed the code according to the guidelines.
But in others, it was simply a matter of the varying tolerance built into different scanners. In these cases, retailers had batches of products that either had to have a new bar code label stuck over each original label or had to be returned to manufacturers.

Finally some products which passed that test, had variable quality printing which meant that some, but not all, could be read at a checkout. In some stores, a system was set up whereby checkout operators recorded these numbers so that manufacturers could be informed and quality improved.

My perception as a customer is that this obsession with non-reads and the zero-tolerance approach has relaxed to the extent that I have not observed a checkout operator noting such products for many years. Instead, I see on many occasions the checkout operator resorting to key entry of the number under the unreadable bar code.

Of course, today's scanners are far more tolerant and are able to read a much greater area of packaging successfully meaning that non-read incidence is reduced. So perhaps this has led to a retreat from the early obsession with printing and reading quality.

But these things are relative. The commitment to quality and to the 99.99% successful read rate specified by the pioneers in their first specifications so many years ago stands comparison with any other area of products and services in my daily life. So, whilst noting that occasional codes are badly printed and some labels are crinkled, we should applaud the originators and developers of the system for its astonishing resilience.

And that must be enough about these lines. It is time to move on from a general understanding of item numbers and bar codes to the practical experience when they finally appeared at the checkouts of retail stores.

12. EPOS Is Not A Greek Island

The presentation I gave in 1986 was to a gathering of journalists and major investors in London. I can still recall my opening slide (before the days of Powerpoint presentations) which showed a picture of a sun-kissed beach with the caption, *'EPOS is not a Greek Island'.* The number of stores in the UK at the time was very limited and the subjects of bar coding and Electronic Point of Sale (EPOS) were not widely understood. EPOS had not entered either the general consciousness or the dictionaries. My explanations were well received.

That cannot be said for my 'words of wisdom' delivered the following year. For in 1987 I was on the receiving end of two difficult interviews, one on radio and the other on television. The general subject on both occasions was the new supermarket checkouts with their scanners and their bar codes. The particular subject was the reaction of customers to their introduction. I had been chosen to be the retail industry representative because I had been responsible for the introduction of scanning to Boots the Chemists a few months earlier and was also chairman of the EAN organisation in the UK.

I knew where the attack would come from, but that did not help much. And I was right. I sat as part of the television audience whilst the microphone roved around disgruntled customers. Out came complaints of products which had no bar code on them, bar codes which would not scan, checkout operators who did not know what they were doing, and ever longer queues at checkouts. All of that formed the opening salvo before the real onslaught began, the barrage of complaints because price tickets had been removed from products in this new world of bar codes. The presenter moved up the gangway steps towards me, clearly pleased with the show so far and looking forward to the next couple of minutes. *'And what do you say about that, Mr Berry?'*

I cannot recall the detail of the subsequent interview but do remember questions like, '*Do you really mean to say ....*', and '*You've heard what's been said, so why don't you ...*'. Needless to say, her summary at the end of the programme did not project the wonders of this new technology.

As you shop not only in supermarkets but in most other retail outlets, you accept the world of bar codes and scanning as a normal way of doing business which is generally very efficient. It provokes few complaints and little anger. It was not always so. For the pioneers in the use of this technology, the scars remain.

*****

During 1988, as scanning systems moved into more chains and more individual stores, concerns resulting from the experience of customers continued. There were many subjects of complaint but the most publicity was given to pricing errors. It was this that was seized upon by consumer watchdogs, politicians and the media.

On 4th October 1988, the front page headline of the Daily Mirror in the UK was devoted to the subject. '*The Great DIY Store Rip-Off*'. The leading article claimed that, '*A check on one store revealed that it was overcharging on 45 per cent of its goods stamped with computer bar codes ...... Customers not bothering to check their receipts would not know they were being cheated...... Hounslow Council has now asked Trade and Industry Secretary Lord Young to bring in strict controls to prevent bar code overcharging.*'

In February 1989, the Daily Mail devoted its front page to '*More Errors At The Till*'. Its story told of overcharging by a supermarket chain and the response of the Consumer Affairs Minister, '*I think what is happening is absolutely scandalous. It cannot go on.*'

I have cited only two examples in one country. I could have chosen countless others from many countries around the world. So what was happening in retail chains at this time to cause such adverse publicity? And what is different today such that similar stories rarely, if ever, appear in the press?

I was expanding the use of scanning into Boots stores at this time. Starting with one store in 1986 and a second some months later, Boots embarked on an installation programme for over 1000 stores, which by 1988 involved converting one store every week to the new system. The hardware and software challenges were immense. But more significantly still, thousands of store staff were involved in learning a completely new way of working. The success and accuracy of the system was dependent upon their actions and getting all of these new procedures correct quickly. We were particularly aware of the difficulties in this environment of ensuring that every shelf-edge price matched every price that was charged at the checkout. But we also knew that the system had far fewer errors than previous systems which relied on cash register staff accurately key-entering prices from price labels.

Despite our best efforts, we were not immune from error as the Surrey Comet headline reported in June 1989. *'Store fined for blunder in prices'*. The case involved *'packets of plastic knives, forks, teaspoons and paper plates'* at the Boots Kingston store. A trading standards inspector *'posed as a shopper'* and *'it was noticed that the price flashed up on the till was at least 10p higher'*. The Boots lawyer accepted the *'isolated examples of human error'* and observed that *'as soon as the mistake was discovered, all 60,000 products at the Kingston branch were checked, and six other mistakes were found – only one of which was to the customer's disadvantage'*. Boots was fined £1,000 and ordered to pay £150 costs.

I continued to be the chairman of the UK EAN organisation at this time. Therefore I knew of the efforts being made by all retailers to reduce pricing errors. But as Nigel Whittaker of Woolworths said in the Financial Times of media coverage at the time, *'It is ironic that the introduction of EPOS tills has substantially reduced the likelihood of till errors, and increased the likelihood that customers will spot them'*. Andrew Osborne, as the general secretary of UK EAN, spent a lot of time and nervous energy dealing with these reports, consistently fending off demands for more legislation and quietly repeating the advantages of the new systems.

And, he observed that when errors are made,' *The Trade Descriptions Act is there to stop stores charging higher prices than are displayed on shelves and customers should enforce their powers'.*

The most damaging demands at this time were for the reintroduction of individual product price marking. If that had been enforced, stores would have had to ensure each computer price matched every individual product's price, not simply the shelf edge price of the product, leading to far more errors. In some states in the United States such legislation was enacted. Fortunately in the UK and almost all other countries, no such action was taken.

From the outset, the pioneers realised that the removal of price tickets from individual products would be contentious. But they were sure that it was necessary. Of course, the first consideration was one of cost. The existing systems of managing paper-based price lists in stores, finding the correct price and attaching price tickets, often using pricing guns, to every unit of every product was very labour intensive, very expensive and error-prone. An early study in the United States suggested that about 20% of the savings from scanning systems would come from the removal of this operation of individual price marking.

But there was more to it than that. The new automatic pricing system would not be subject to the same price errors because it would only rely on the correct price being set up on a central computer database and transmitted to stores. All other human errors would be eliminated. It was a far better and far more accurate system. But if retailers were forced to operate both systems at the same time by retaining price stickers on products, there would be endless challenges because any manual error would mean that the correct scanned price would not match that on the individual price ticket. So both the number of problems and the cost of the overall system would increase rather than decrease. It was clear that the removal of price tickets was not only financially desirable, it was practically desirable too.

These events happened over twenty years ago. This raises the question, *'What is different today such that similar stories rarely, if ever, appear in the press?'*

The simple answer must be that retailers have over twenty years of experience using the systems. In addition retail staff are operating with methods they are familiar with rather than working in a period of major upheaval and change. Refined and established procedures, checks and training methods have ensured that accuracy is now very high, giving customers confidence in the systems. The early criticisms and threats to enforce individual price-marking did result in a lot of attention being given to those circumstances where errors were most likely to occur. Though trading standards officers and others continued to unearth errors, they were aware of the improvements that had been made and recognised that some human mistakes were bound to occur.

But there is a second answer too. Bar code scanning has become more widespread and is an accepted fact of everyday life. As a result, customers no longer focus on this issue, no longer check their till receipts to the same extent and no longer challenge retailers. Perhaps that is because their personal experience tells them that errors do not occur.
However, my own experience suggests that, when I do check, a small number of mistakes continue to be made. Even if my findings are unusual, I suggest that complacency is dangerous for both customers and retailers.

It is possible that for you, your partner or your children the supermarket shop does not represent the high spot of your week. So perhaps some mild entertainment can be added by giving one of you the responsibility of price auditor on a future shopping trip. This role entails writing on a notepad the shelf edge price found for each item purchased and the price per kilo of any loose products bought, whilst noting any missing shelf edge price labels. On unpacking, it means checking off all these prices against the till receipt and identifying any mistakes. Finally, if you have the energy and time, it involves writing to the retailer if any errors are found.

I hope you will find it a waste of time because there are no errors, and be reassured that the removal of individual price marking when article numbering and bar coding were introduced has been justified. But if that is not the case, your price auditor will have earned some money, and the retailer concerned may make renewed efforts to ensure that the scanning system works as intended.

*****

The fundamental design of the system (a unique non-significant number, an easily-readable bar code, a price-look-up) is so simple. And it is that simplicity that ensures it is capable of being managed and operated in so many countries, in so many stores, and by so many retailers and manufacturers. But do not be deceived by that simplicity.

As I have noted, Boots introduced its first scanning store in 1986. The newly developed store system was supplied by IBM after the most exhaustive hardware and software testing by them in the US and the UK. But for Boots the failure of the system in a store could have resulted in the temporary closure of the store, the loss of business and the loss of both staff and customer confidence in the new system. So we decided that we must superimpose our own testing regime on that of IBM. Our own software developers listed all the errors they found - they stopped counting after the testing found the *one-thousandth* software bug, but they did not stop testing. This is not a criticism of IBM but rather an indication of how many different circumstances the computer system had to deal with from millions of customers and millions of products. No, it is not as easy as it seems.

Those responsible for the development of what you see today went to enormous lengths to ensure robust systems. They had no option. Customers were not interested in how difficult it was to introduce the new scanning technology, how much testing we had done, and how good our system was - as I soon learnt.

In September 1986, we were proud of what we had achieved. The Boots store in Peterborough had been converted to scanning the previous month. EPOS had moved from a few food supermarket chains to UK general merchandise stores in Boots and WHSmith. We knew how much effort had gone into this, how much had been achieved and we were right to be proud of it.

When the decision had been taken in principle to move in that direction two years earlier, not only was there no practical computer solution to achieve it, but also almost none of our suppliers knew much about bar codes. We were lucky.

In that period, the power and capacity of personal computers had increased to the point where they were capable of operating a store. And improvements in networks allowing data transmission within in a store and between a store and Head Office meant that operating EPOS was possible.

Even so, it took a large team of skilled people to achieve the opening date in August. And, despite giving all our suppliers 18 months warning of our plans, only about two-thirds of our products were bar coded by that date. A back-store operation was necessary to add bar code labels to the remainder.

But the change was made with no major incident. So as I went round talking to the staff at Peterborough, I was pleased that the system had been accepted by staff and customers, and that we could concentrate on the planned programme of conversions to EPOS in other stores.

It was a sales assistant who made me think again. When I asked her if she was happy working with the system, she gave me the positive reply I had expected. But I then asked her if there was any change we could make to the system to make her life happier. Immediately she said that we should print the till receipts more quickly.

This was not a reply that I expected. As I explained to her, we knew the importance of the new till receipt with descriptions and prices, so we had designed the fastest printing of a till receipt by any EPOS system in the UK. I can still picture what followed :

She picked up a waste bin under the till and said, *'Then why are all these till receipts here? It's because customers can't be bothered waiting for them to be printed. So they go off without them, sometimes complaining to me, and I throw them in the bin.'*

I had not heard of this problem from any supermarket chain. Perhaps it was because we had many customers who bought only a few products during lunchtimes or after work and who wanted speedy service. Whatever the reasons, it was clearly a problem for some customers and, as a result, a source of stress for our staff. So despite having the fastest system in the UK, we knew the system must be changed and it was redesigned to reduce the time it took to print the till receipt by 25% before we converted the next store.

I have not mentioned this example or the experience of Boots because it was exceptional within the retail industry. The introduction of EPOS and the adverse pricing publicity reminded all retailers of the long-standing mantra that 'retail is detail'. They were rightly expected to be accurate and reliable in terms of every customer, every transaction, every product and every price. Robust standards of article numbering and bar coding were only a starting point. Robust computer systems and operational procedures were essential too.

The world of the 1980s and of media outrage at the so-called rip-off by retailers when introducing these early EPOS systems seems a lifetime away. Because scanning is now so widespread and has been in use for so long, it might be assumed that everybody understands what is happening in the world of EPOS. But my ongoing conversations on the subject indicate that this is not the case.

So I will summarise what happens within a retail store in the next chapter, and in particular detail how prices are managed and the pricing problems referred to in this chapter avoided.

13. Beep Beep

Our starting point is the retail checkout. The checkout operator is presented with a product containing an article number and bar code. As we have seen, the number is unique to that particular item and the bar code is a machine-readable representation of the number. In order to complete the sale of that product, the operator only really needs one other piece of information, its price. However, as we saw earlier, it is not contained in the number or the bar code. So where is the price found?

A computer file holds all the article numbers of products sold in the store and the associated price of each. This file is usually found in a computer in the back of the store, but it can be in the till itself or theoretically could even be outside the store and linked to the till by wired or wireless connection. So when the scanner reads the bar code, it is translated into the article number and the computer file is accessed to obtain the price of the product. If for some reason the bar code cannot be read, the checkout operator can key in the printed article number directly and retrieve the price of the product as before.

But you will be used to more sophisticated systems than this. You would normally expect that when the article number is found on the computer file, it will have not only the price but also a description of the product that can be displayed at the checkout and printed on the till receipt. All major supermarket systems provided this from their first scanning stores, pointing out to customers that one of the benefits of the new system was that it gave a more detailed till receipt than had been possible before. However, it was also a necessity. Previously a customer could match the prices on a till receipt with the price tickets on products to check that they had been charged correctly. Now, except for those products which had a pre-printed price on the packaging, there were no price tickets on products and checking the till receipt would have been almost impossible without product descriptions.

That is the simple process by which the price of each item is found. Of course it relies on an accurate computer file which contains the current price of every product. But that requires much less error-prone human activity than the earlier system of attaching a correct price sticker to each individual item found in the store and entering that price at a cash register. However, the removal of price stickers means that a customer cannot find the price on the packaging of most products. And so an associated operation is necessary which provides the price found on the computer file as a shelf-edge label which can be found by a customer at the point when the decision whether to buy is made. These labels are printed from the computer file and so the price charged to the customer at the checkout should always be the same as that shown on the shelf-edge label.

*****

But how do retailers ensure that shelf-edge prices always match computer prices? And, in particular, how do they deal with the change of a product's price when errors are most likely to occur? It was noted earlier that altering the price of a product in a particular store does not mean that a new article number and bar code are necessary for the product.

So for such a 'normal' price change of an existing product, both the computer price and all associated shelf-edge prices relating to this unique article number must be changed. But since the price on the computer cannot be changed at precisely the same moment as the shelf-edge label is physically altered, how can a system be devised which ensures that the customer is always charged the price found on the shelf-edge when the item is scanned at the checkout?

The designers of the system recognised this problem and devised a simple but necessary guideline for such price changes. Computer and shelf-edge prices should always be changed in such a way that the customer is never charged more than they expect from the shelf-edge label. If any pricing differences occur, they should always be resolved in the customer's favour. What does this mean in practice for the timing of price changes made in the store?

Of course the simple way to manage the process would be to make all changes to the computer price file and shelf-edge prices when a store is closed. But that is not always possible, particularly when stores are open for 24 hours. So what happens if changes are made during trading hours?

If the price of a product is increased, the shelf edge price must be altered before the price change becomes effective in the computer system which provides the scanned checkout price. And there must be a sufficient time delay to ensure that a customer seeing the lower original shelf-edge price will have passed through a checkout before the higher new computer price becomes effective and is charged when the bar code is scanned. Of course that means that in a few cases the customer may see the newly inserted higher shelf edge price but reach the checkout before it is effective in the computer file and thus be charged the lower original price.

On the other hand, if the price of a product is reduced, the shelf-edge price must be changed after the computer price, even if this means that some customers read the higher original shelf-edge price but find that they have been charged the lower new price when the bar code is scanned. In other words, the customer should always win. This 'best practice' is a way of seeking to ensure that human errors do not result in an incorrect higher price being charged to a customer.

Having dealt with the simple and efficient procedures for what happens when there is a 'normal' price change, it is necessary to look at what happens in slightly more complicated instances where some form of price promotion takes place. These can take a number of different forms.

- The simplest is where the price of an item is reduced for a limited period of time as a special offer. In this case, like the 'normal' price change, provided that the price on the computer file associated with this article number is changed, the lower promotional price will be retrieved from the file, displayed at the till and printed on the receipt.

When the promotion ends, the price on the computer is changed back to the normal price. So the product does not need a different article number and different packaging during the promotional period. The same item can be on a shelf before, during and after a price promotion, and the correct price can be charged from the same article number, because the current price is maintained correctly on the computer file.

Of course, shelf-edge prices must be changed at the beginning and end of the promotion to match the computer price.

- However, if a <u>special pack</u> of the product is produced with a price printed on the packaging, then both the original pack and the promotional pack could be in a store at the same time. Since only one price can be associated with one article number on the store computer, this special pack must have a different article number with a different computer price. In a similar way, a pack which offers an extra quantity (say 100% extra free) is a special pack, a different pack to the original and so requires a different article number. For example :

Weetabix – pack of 12   5010029000504
Weetabix – pack of 12 price mark £1.50   5010029215588
Weetabix – 24 for price of 12   5010029210200

- What about <u>multipacks,</u> where several of an item are bound together offering a lower price than when the same items are bought separately? Again, this pack is different and must have a different article number, bar code and price. In the early days of scanning a few of these products caused problems because, although a multipack was given a new article number and bar code, the original bar code of the individual item could sometimes be seen through the multipack packaging. So the scanner could read that number by mistake and the till recorded the sale price as that of the single item rather than the multipack – good news for the customer, but retailers were not amused.

When you buy a multipack today, you will find either that the individual product bar code is completely hidden by the packaging or possibly that the manufacturer has produced individual products within the multipack which do not have any bar code on at all. A multipack example is :

JW Tuna Chunks in Spring Water – 185g   5000232812438
JW Tuna Chunks in Spring Water – 4x185g 5000232812445

- Buy one get one free. Now we enter a more difficult world which cannot be dealt with by changing the article number. The first customer may buy only one of the product and the price will be found and charged in the normal way. However, the next customer may buy two of the product, and since both have the same bar code, the same price is found for each. So a separate piece of computer software is required to identify these products, to notice when the second one is scanned (even if it is not immediately after the first one), to override the normal price and give the product away for free. This explains why the printed till receipt normally shows the price of each item as the code is scanned but then prints afterwards the discount given from 'buy one get one free'. There are many variations of this type of special offer where no new article number is required but the price is determined by the computer's table of price promotions. For example, if a price promotion offers cans of HP Baked Beans normally priced at 44p at the special price of 3 for £1, the scanner will read the bar code on each can, and the computer system will produce a charge of 44p each time but give the discount of 32p after the third is read.

- Buy one product, get a different product half price. This is similar to the last case and once again is dealt with by a separate piece of computer software but without changing any article numbers. Both products are scanned, the article numbers lead to their normal prices, but the software identifies them and gives the discount (even if the discounted price item is read before the primary item).

- Reductions to clear are dealt with in two different ways depending upon the type of reduction. If all the stock of a product is reduced, then the computer file price can be changed to the lower price, as in the case of the earlier price promotion. However, if only some units of the product are reduced, for example items close to their sell-by date or damaged stock, a different solution must be found because there would be two different prices for the same product (with the same bar code) in the store. The computer file price for the product cannot be changed, for some 'normal stock' of the product will be sold at the full price and some 'dated or damaged' stock at the reduced price in the same store. So the key here is that a reduced price sticker placed on some units must cover the bar code. In this way, whereas full-price stock can be dealt with by the normal reading of the bar code, reduced-price stock can never use the obscured bar code. Such products are dealt with in one of two ways. In the first, the operator must enter the reduced price from the price sticker and override the normal price-look-up, before peeling it off and reading the bar code to give the product description (but not the original price) and record that a sale has been made at the reduced price. In recent years an alternative method has been developed to avoid peeling off the sticker, which will be dealt with in the next chapter.

For the first scanning stores, the size and speed of computers presented a great challenge in terms of how much information could be stored and processed. But nowadays these limitations do not apply and marketing executives can make ever greater demands on software programmers to support their latest 'great idea' to give extra sales.

So, in summary, most of the products passing through supermarket checkouts fit into one of three groups :

- Scanned bar code only – leads to the price held on the computer for normal price merchandise and simple special offers.

140

- Scanned bar codes and table of more complex special offers – leads to the appropriate price reduction.
- Override price and scanned bar code – for reductions to clear, say, of out-of-date stock.

But before we leave the basket of goods presented to the check-out operator, we must deal with one other type of product which does not always fit into any of these three groups.

*****

All of the products considered so far in Part II have one fundamental feature in common. They are 'fixed'. In other words, each one of them is produced with its fixed quantity, weight, volume etc – for example, a carton of 6 medium eggs, a 400g tin of tomatoes, a 1 litre jar of apple juice. Each has its own unique article number which applies to every unit produced of that particular product. And each will have an associated price held in the computer file and displayed in the store where it is sold.

Of course, that price can be different in different stores, for each store has its own computer file of article numbers and prices. And, as we have seen, it may be sold as a special offer with a reduced price for the period of that offer. But the product and its number are 'fixed'. And if the product changes in any significant way, a new number is required.

This fixed number system also applies to such products as fruit, vegetables, meat, cheese etc if they are in standard weight packs ; for example, 1kg of Granny Smith Apples, 5kg of Maris Piper Potatoes, 500g of Mature Cheddar Cheese. But what if products are sold in packs, but those packs are of differing weights? And what if items such as vegetables, cheese and fish are sold loose and, when put in a shopping basket, also have variable weight? These products cannot have a fixed article number with associated fixed price. They require a different solution to that which has been described so far.

141

Since the bar code of a pre-packaged product with variable weight can be scanned at the checkout without being weighed, it implies that it must have information about either the weight or the actual price of that particular pack. Similarly, if a loose product is weighed at a service counter and a bar coded label attached, that must include such data. In many stores, particularly in continental Europe, customers can weigh their loose products with a bar coded label being printed for them to attach. Unlike all the codes and numbers we have considered so far, these do have specific information about price included within them. How that is done will be explained in the next chapter.

Some loose products may fall outside this system and be weighed at the checkout, as happens at the shops in my town. In this case no bar code is usually involved and the price is found from a computer file at the checkout.

But having moved beyond the most common article numbers to consider a different group of products requiring a different solution, you may have your own questions to raise. So anticipating that I have headed the next chapter, *'But what about ...'*

14. But What About ...?

I can vividly recall a question posed by a friend at university fifty years ago. Why, he asked, do people misuse the expression, 'The exception proves the rule'? Not expecting or wanting an answer from me, he went on, 'It is always quoted when a rule is broken to suggest that somehow this validates the rule. But that is nonsense. The opposite is true. It is when an exceptional circumstance still conforms to a rule that you can say that the rule is proved.'

I remember that now as I consider rules and exceptions. The first chapters in this section introduced article numbers, their associated bar codes and the checkout operation. I hope that has provided a broad understanding of the subject. However, I have not told the whole story. The basis of the system shown to date has been :

- Globally Unique article numbers given to products
- Unintelligent article numbers with no embedded meaning
- Article Numbers together with Bar Codes that represent them
- Different article numbers for different products
- All numbers conforming to an 8, 12 or 13 digit standard

This is the basis of the system but they are not rules. Your understanding of the system will now be widened as we introduce exceptions to each of these five statements. For you may have noticed that not all the numbers that you will find on products in store and in your home conform to the detailed structure of the numbers described earlier. You may be saying, '*But what about .....?*' If so, the exceptions in the next two chapters should answer those questions.

*****

The first exception will address Global Uniqueness.

Going back to 8-digit GTINs, the examples quoted earlier were :

| Baby Bio Plant Feed | 175ml | 503 7128 0 |
| Schwartz Coriander Leaf | 6g | 509 2631 2 |

These had a number structure made up of a GS1 country number followed by a GS1 (not company) allocated number and check digit. This was necessary because the total of 8-digit numbers was limited.

However, the first group of products in front of me have an 8-digit number and bar code, but they do not conform to the format set out earlier, because the first two or three digits do not represent any GS1 Country Prefix found in the list of two and three-digit country prefixes.

Boots Salon System – Cond.	300ml	02465722
Sainsburys Sunflower Oil	1 ltr	00064491
B&Q Organic Bone Meal	5kg	03074763
WHSmith Notepad		00522618

In looking at them, you may note that there is something that unites both the <u>products</u> and the <u>numbers</u> which they have been given. All of them are 'own brand' products, produced for a specific retail chain and all start with the number zero.

There was a particular historical reason why they were introduced. Before the introduction of scanning some retailers used short 'velocity codes' for some of their best selling items. These short numbers could be key-entered at a cash register to find the price, thus avoiding the need to attach a price ticket to every single pack. The zero-number system allowed such retailers to retain those velocity codes if they wished within the new system. There was also a logical reason for treating own brand products differently. Whereas the basis of the EAN/UPC system was that every product needed a unique global number so that it could be sold anywhere, own brand products did not need such numbers because they would never be sold outside a particular retail chain. In other words, if Sainsbury's gave the number 00064491 to one of its products, it was necessary to ensure that it did not give that number to any other product it sold.

However, it did not matter if another retailer used the same number, provided the product was only sold in that retailer's stores where the number could not be duplicated and cause confusion.

In fact 'own brand products' is not the best terminology. These numbers could apply to any product from a manufacturer which is uniquely made available for sale in only one retail company. So this group of numbers is referred to as Restricted Circulation Numbers and the specific number system as RCN-8. Restricted Circulation Numbers always begin with either 0 or 2. The user then forms an 8-digit number by adding their own product number and finally calculating the check digit as before. In order for these numbers to be recognised, no country prefix begins with either 0 or 2.

Two benefits became apparent from this decision. Firstly, because many products did not need a globally unique number, the pressure on number capacity of the shorter GTIN-8 numbers in the future was reduced. And secondly, those using these numbers could use shorter 8-digit numbers rather than 13-digit numbers on larger products if they wished (eg Sainsbury's Sunflower Oil), whereas GS1 country-allocated short numbers were limited and therefore restricted to smaller products with limited space for printing.

The major downside for a retailer or other user of these numbers has always been that if, in the future, they decide to sell their products in other stores, then because the numbers may not be globally unique, each of these products would need to be renumbered with a unique global number. Perhaps because of this, some retailers have chosen not to use these numbers but rather to number their own-brand products in a globally-unique way. They have applied for their own company number and added their own product number, as shown below :

Coop Baked Beans	500 0128 28574 2
Somerfield Baked Beans	500 0192 15545 3
Tesco Baked Beans	501 8374 23071 3
Morrisons Org. Chop. Tom. 400Gm	501 0251 23453 1

I should note that a further downside to using non-unique numbers has emerged more recently. The use of the article number has widened and, for example, it may now be read on a mobile phone to access information about the product. But if that number is not unique and can apply to different products in different retail chains, information might be retrieved for a different product to the one being scanned.

So, our original answer to the question, *'What's In A Number?'* was that each number is unintelligent and globally-unique. We must modify to say that this particular group of numbers are also unintelligent but may not be globally-unique, because they do not need to be.

*****

We must change our answer again for the next group of products I have in front of me and their article numbers. Here we need to say that they are Intelligent Numbers, because they need to be.

One of the key principles on which the EAN/UPC system was based was that the article number should not include the price of the product. As we have noted, to have done so would have implied that a different number was required whenever the price changed. But, as we saw in the last chapter, for a particular group of variable weight products, this principle presented a problem.

Take this example. A store is selling rump steak, which is cut, packaged and sold according to its weight. But since the weight of each pack is different, the price of each pack is different. Clearly, every pack cannot be allocated a unique number, so what options were available?

One was to give a number to the product 'rump steak' and to add a new extra number and code which gave the price of the particular pack. Then the scanner could read the main code to obtain the product details (without price) and read the add-on code for the price.

146

Of course, that would mean that scanners would need the additional capability of reading add-on codes. The alternative was to find a way by which a code and price could be included within the 12 or 13-digit code for these specific types of product. This latter solution clearly had advantages and a solution was developed to deal with instances of both branded products and in-store own brand products.

For branded products, it was decided that the EAN (now GS1) country prefix would be replaced by any of the numbers 20 to 29, though not all countries use all of that range. These numbers are referred to as RCN-13 rather than GTIN-13 and have a different structure. Computers could be programmed to recognise that 13-digit numbers beginning with some or all of this range are distinctive 'abnormal' variable weight product with the price included in the number. For own-brand products, any 13-digit number starting with 02 would be recognised in the same way. These are referred to as RCN-12 or UPC prefix 2. The codes and numbers for these products are unlike those considered earlier. All previous codes have been read and the number found on the database to obtain its price. But these codes are produced for each pack, have no number on the database and find the price within the code.

The structure of each of these codes was agreed in 1977, but a problem was identified which meant that a single international standard could not be agreed immediately. That problem was national currencies. The nature of each currency meant that more digits were required to express a price in some countries than others. Even for Europe alone, in those pre-Euro days, the difference between, say, the UK pound and the Italian lira was marked. And the more digits that were required for the price, the fewer were available for the product number itself. The solution was a simple one. Because sales in a particular country would always be made in a particular currency, each country would decide on a national standard for the detailed length of the price within the broad international structure of these codes.

Given that basis for the codes, these are the current standards in the UK :

RCN-13   20 NNN NN K PPPP C
RCN-12   02 NNNN K PPPPP C

In the first case, NNNNN is split into a three-digit company prefix assigned by GS1 UK and a two-digit item reference given by the company. As a result, only four digits are available for the price and the maximum is reduced to £99.99.

In the second case, NNNN is the retailer's own item reference number (max 9999) , PPPPP is the price (max £999.99), C is the normal check digit, and K is a price check digit.

Some explanation of the price check digit K is necessary. As we have seen, the general check digit C is effective for the vast majority of errors which may occur. Even in the few instances where an error does produce the wrong number, if that number is not on the product database it will be rejected. But for the group of items where the price is included, no number exists on a product database. Therefore, it was decided from the outset that the inclusion of a price must be accompanied by a more complete check such that price reading errors could never occur.

A separate check digit K is derived from a formula based on the price digits alone and is calculated to complete the 12-digit number, followed in the normal way by the calculation of the 13th digit, the general check digit. In other words, a double whammy! For those interested in how the check digit K is determined, the detailed calculation is included in an Appendix.

Here are some examples of some RCN-13 numbers :

Coop Chicken Small - £2.99 per kg	02 9118 0 00374 9
Price £3.74	
Coop Chicken Large - £2.23 per kg	02 9114 2 00421 0
Price £4.21	
Tesco Rump Steak	02 3130 4 00356 9
Price £3.56	
Somerfield Cheese	02 9963 5 00463 1
Price £4.63	

When the Somerfield Cheese is packaged :

- The cheese-packer weighs the cheese and enters the item reference number 9963
- The computer system finds the price per kg for this product and calculates the price of this weight of cheese 00463
- The system then calculates for the price 00463 the price check digit 5
- The system then calculates for the 12-digit number 029963500463 the general check digit 1
- The attached printer then produces a label with the number and bar code 0299635004631.

At the checkout, the label is read, the general check digit is verified, the specific prefix 02 is recognised, the price check digit is verified, the product description is found from the item reference number in digits 3-6 and the price in digits 8-12 is found and printed. And of course, that is done in less time than it has taken to describe it here.

*****

Your understanding of article numbers and their associated bar codes now encompasses unintelligent globally-unique, unintelligent restricted circulation and intelligent price-sensitive numbers.

However, all are linked by the fundamental decision made when the system was developed that the article number and bar code were inseparable.

But how do you get a readable 8-digit number and bar code on a lipstick or an eyeliner? In 1985, as we prepared for the first Boots scanning store, we were the first retailer in the world who was faced with the problems presented by the EAN/UPC system for very small items. Our interest was not only in cosmetics but also in pharmaceutical items, many of which were also either small or had very limited space on packaging.

We decided that it could not be done. And so we brought to EAN a key issue. Although the basis of the system was a unique global product number with a machine-readable symbol representing the number, if there was insufficient space to print both, surely it was better to print only the number rather than exclude such items from the system. The response was mixed. Some were sympathetic and supported our stance. Some felt that bar codes had been introduced to serve customers more quickly and that key-entry of numbers should only be used where a code would not scan. They argued that manufacturers should find a way to include a bar code, even if it meant packaging products. However, it was finally agreed that a Number-Only solution could be used as a last resort, if no other method for including the code could be found.

We still had a problem. The only label on many lipsticks at that time was found as a small circular one stuck on the end. Was there even room for a number and, even if it could be printed, would it be readable? Our earliest solution was to print the number and provide a small magnifying glass at the till to read it!

Oh, how times have changed. When writing this chapter, I visited a Boots store to check whether any magnifying glasses still existed. They do not. I did find one product range where the number was still printed on the end of the lipstick.

But for all other ranges, a printed label was stuck onto the side of the lipstick with not only a bar code but even with a 13-digit number. So, what has changed in that period? Several things have come together.

The cost of printing and attaching a label has reduced, scanners can now read much smaller bar codes, retailers want to avoid number-only solutions even more than in the past and finally, cosmetics manufacturers realise that the printed label does not cheapen or detract from the product in the eyes of their customers.

However the effect of this historical debate has been that you may still find some products which are small or have limited printing space on packaging which are part of the GS1 number set but have no printed bar code.

*****

Greeting Cards do not have the 'small item' problem seen in some cosmetic products. So there should be no difficulties in printing unique numbers and bar codes on these products, should there? Of course, as with the cosmetics manufacturers and many other producers, there have been concerns that these codes would detract from the appearance of the item being sold. But they posed a different challenge to the EAN/UPC system.

When scanning was introduced to WHSmith and Boots in the mid-1980s, the question was raised as to whether it really was necessary for every product to have a unique article number, a principle of the system. It was noted that a card supplier could be producing hundreds of different designs, sold at the same price, which had a limited life before being replaced by different designs. Surely, in these circumstances it would be better to have a Product Range Number rather than the options of hundreds of different numbers or leaving them all out of the new system? That would save not only a lot of administration costs but also number capacity.

The issue was not confined to greeting cards. On a far bigger scale, the same problem was present in the clothing industry. A particular design of shirt, say, might be available for sale in a range of sizes and a range of colours, with all sold at the same price and only available for a limited season. In this fashion world, was a different approach required to the numbering of individual products and ranges of products?

Once again the response was mixed. Some felt this was a pragmatic solution, others that it compromised the whole system. Finally, the principle of unique numbering of all products was firmly reiterated, though it was accepted that, if retailers and manufacturers agreed, there might be particular circumstances in which this was relaxed. As with a number of decisions taken, it was a practical solution at that time.

So, when looking for Christmas Cards, I found the following :

Card Design 1 – Niece (Price band J)	502 173490094 2
Card Design 1 – Nephew (Price Band J)	502 173490094 2
Card Design 2 – Special Friend (Price Band K)	502 173490071 3
Card Design 2 – Granddaughter (Price Band K)	502 173490071 3

In other words the card manufacturer had grouped a number of different cards according to their price, still allowing the retailer to find that price by reading the bar code.

But once again, times have changed and some card and clothing manufacturers have not grouped similar items. For example :

BHS Shirt – Size 16 – Red & White	50 515249 1033 2
BHS Shirt – Size 14 – Red & White	50 515249 1032 5
BHS Shirt – Size 14 – Green & White	50 515249 1039 4
BHS Shirt – Size 14 – Brown & White	50 515249 1053 0

So, although you may see some examples of 'group coding' today, the world has moved on and many of these products are now sold with unique numbers.

The major reason why a card or clothing manufacturer and retailers have decided that each product requires a unique number rather than a group number will become apparent in Part III, when the wider effects of this numbering and coding system on the whole retail industry are considered.

*****

When planning this part of the book, I thought about the people who had asked me about article numbers and bar codes over the years. In trying to answer their questions, I always thought of the products in the shopping trolley or shopping basket which were presented at checkout or service till. As I mentally unloaded these products to explain the system, I must confess that in many cases the glazed expression in the eyes often told me that I had travelled a bar code too far. But for others, their interest, or perhaps mere politeness, led them to ask as I paused, *'But what about ....?'*. And for some, a rare breed indeed, my answers were followed by a further question, *'Yes, but what about ....?'*.

So, unable to see your eyes but knowing that you have reached this far, I am hoping and assuming that you are part of this select group. If so, there is only one other group of products I have planned to write about. These are products where it was considered necessary to say that they should not be confined to 8, 12 or 13 digits but required additional information that required an extra add-on code to form a <u>Longer Code.</u> And these form the basis of the final chapter of this second part of this book before I tie up some loose ends and summarise what has been achieved in stores at checkouts by this retail revolution.

15. Yes, But What About ...?

When my youngest daughter left home, I assumed that all of the contents of her room would go with her. Such has not been the case. Of course one benefit has been that when she comes back to stay for a weekend, she has many of her possessions around her. An unexpected second benefit has been that I have been able to raid her room to illustrate the next example of article numbers and bar codes.

Among the books on her bookshelves are two of the most popular titles ever printed, one relatively popular book on the subject of global warming, and one which, how shall I put it, has somewhat limited appeal and sales.

Harry Potter and the Order of the Phoenix	9780747551003
Harry Potter and the Half-Blood Prince	9780747581086
Heat by George Monbiot	9780713999235
On the Duty of Man and Citizen by Pufendorf	9780521359801

All have the standard 13-digit article number with associated bar code that you have come to know and love. And all have a similar format of 3-digit prefix, 9-digit number (made up of publisher and publication) and check digit, though instead of the first three digits representing a country organisation, all have the GS1 prefix 978, which has been given to the world book-numbering organisation.

However, all of them have something printed above the bar code starting with the letters ISBN. I have listed below the 13-digit GTIN and the ISBN printed for each book. The numbers are similar but not identical.

GTIN	978 074755100 3		ISBN	0-7475-5100-6
GTIN	978 074758108 6		ISBN	0-7475-8108-8
GTIN	978 071399923 5		ISBN	0-713-99923-3
GTIN	978 052135980 1		ISBN	0-521-35980-5

And this brings us to a bit of history. In 1968, a 9-digit Standard Book Number was introduced in the UK. Two years later this was enlarged to become a 10-digit International Standard Book Number (ISBN). The system allowed every book to be identified in booksellers, libraries etc. Furthermore, if a novel, for example, was published in two different versions, say as a hardback and as a paperback, then different numbers were necessary to distinguish them. It was an independent system with its own organisation for assigning numbers. By the 1990s the ISBN organisation recognised that there were two benefits from linking the 10-digit ISBN to the 13-digit article numbering system. Firstly it would mean that bar coded books could be scanned in retail outlets alongside other products and secondly it would give them increased number capacity. For the latter, they were allocated a prefix 978 with an understanding that if the book number capacity became exhausted, they would be given a further 3-digit prefix, effectively doubling available 13-digit numbers. There was only one problem – check digits.

The development of the ISBN was remarkably similar to that of UPC and EAN. The 10-digit ISBN number contained elements of a publisher number and that publisher's own item number followed by a check digit.

But when this 10-digit number was preceded by the prefix 978, the final check digit did not conform to the calculation for the 13-digit number's check digit. The solution found can be seen in the four examples above. The first 9 digits (excluding the check digit) of the ISBN were taken, preceded by the prefix 978 and followed by the calculated 13-digit number check digit. Thus on books you can find both numbers printed. This has allowed the book industry to continue to use its systems of 10-digit numbers whilst allowing others to incorporate books in the global system of longer numbers.

Interestingly the book industry ultimately decided that the future 13-digit article number should also serve as the ISBN.. This meant that the check digit problem and need for two separate but linked numbers disappeared.

The change was made in January 2007, and so going back to my daughter's bookshelf I find that the last book in the Harry Potter series, published in that year, has only a single 13-digit number, as follows :

Harry Potter and the Deathly Hallows   9780747591061

Completing this history of the cooperation between UCC, EAN and ISBN, you will note that some books contain a supplementary number and bar code. For example

Long Walk To Freedom – Nelson Mandela
          9780349106533   00899
The Constant Gardener – John le Carre
          9780340733530   00699

When the book industry introduced the 13-digit system, many countries had retail price maintenance agreements for books. Therefore it made sense not only to identify a particular book number that could be scanned as a bar code but also to print the fixed retail price as a value and associated bar code. In the examples above, the prices found are £8.99 and £6.99. This supplementary or add-on code is still found and used in countries with fixed retail prices. In other countries, the price may be printed but not read, with the actual price charged indicated by the retailer (including on-line retail companies such as Amazon) and contained for that book on the computer file.

*****

Almost all of the books found on my daughter's bookshelves do not contain this supplementary code. But for another group of products, supplementary codes are always found. They are newspapers and magazines.

The Guardian	Wed 21st March	9770261307835	12
The Guardian	Fri 23rd March	9770261307859	12
The Guardian	Wed 28th March	9770261307835	13
Matlock Mercury	Thu 22nd March	9771366144059	12
New Internationalist	April 2007	9773050952124	04

These products are distinguished from books in that they are published in a number of editions having different content. Like the ISBN, a standard numbering system existed for these products – the International Standard Serial Number (ISSN), an 8-digit number with its own check digit. Again like ISBN, this system has adapted to become part of the wider GTIN system.

In this case, the 3-digit prefix 977 was given to ISSN to be followed by the first 7 digits of ISSN (the check digit excluded) followed by a 2-digit sequence variant and the calculated check digit. The examples above are rewritten to reflect this.

The Guardian	Wed 21st March 2007	977 0261307	83	5	12
The Guardian	Fri 23rd March 2007	977 0261307	85	9	12
The Guardian	Wed 28th March 2007	977 0261307	83	5	13
Matlock Mercury	Thu 22nd March 2007	977 1366144	05	9	12
New Internationalist	April 2007	977 3050952	12	4	04

Following the prefix 977, all three editions of The Guardian newspaper have the first 7 digits of its ISSN number 0261307. For the year 2007, the Guardian gave the publications from Monday to Saturday the sequence numbers 81 to 86. So every Wednesday edition of The Guardian had the 2-digit sequence number 83 and every Friday edition 85. Finally the check digit was calculated. This means that the 13-digit GTIN number for any Wednesday edition in 2007 is identical.

The 13-digit bar code is read at checkouts in the normal way and the price found from the computer file. However, the newspaper industry required to track the sales and returns of any particular day's issue. Consequently an add-on code was produced to identify each specific issue for ISSN. In the case above, it is clear that the add-on indicates that these are the 12th and 13th editions of the Wednesday and Friday papers for that particular sequence in the year 2007.

The same format can be observed for the other two publications.

The Matlock Mercury appears weekly and has the sequence number 05 for 2007 and the add-on 12 for the twelfth edition in 2007.

New Internationalist is published monthly and has sequence number 12 for 2007 and add-on 04 for the fourth edition in the year.

*****

You may see even longer numbers and codes on some products. Earlier, I noted that reductions to clear of individual items must ensure that the original bar code cannot be scanned and the full price charged incorrectly. This is usually achieved by a price sticker placed on top of the bar code so that the override lower price can be entered before the sticker is removed and the bar code itself is read. This does slow down the checkout operation, particularly with those customers whose desire for a bargain leads them to search for and buy many such products.

So some retailers use an option which has been developed allowing them to read a new longer bar code stuck on top of the original one which contains all the information required.

An example of this is given below :

Sainsbury's Fruit Trifle   01426717
Sainsbury's Fruit Trifle – reduced to £1.69
                              9100000014267170001694

Looking at the number, you will see that we have moved well beyond any length discussed so far to 22 digits. These digits include both the original article number 01426717 and the reduced price £01.69. In passing, it is interesting to note that in fact two stickers have been added to this product. The first, on the top of the product, for the customer, simply says 'Reduced – Was £2.39 Now £1.69'.

The second, on the base of the product, for the flat-bed scanner, simply gives the long number and bar code stuck across the product bar code on the packaging.

The detailed structure of this number and its associated bar code are not given here. However, it is worth noting that this is only one example of the demands made on the system for particular products or situations by particular retail or manufacturing sectors which have led to the introduction of new standards with differing lengths of numbers and bar codes. This book does not contain an exhaustive list of all that you will find at a retail checkout. So in your shopping trolley you may find other codes which are used by a particular retailer. However, if you require a more complete, more technical description of the many variants of numbers and codes, these can be found via the GS1 global and country websites and organisations.

But it is worth making a general point at this stage. The use of non-significant 8,12 and 13 digit numbers, structured in the way they are, and giving a vast number of combinations is the basis of the GS1 system. However, if more information needs to be read about a product or a transaction, these lengths are a constraint. So theoretically, extra information could be given either by using several numbers or by using longer numbers or by stacking numbers on top of each other. And more will be said on those subjects in Parts III and IV.

*****

Before completing this part of the book, I need to refer to two matters which can be observed at retail checkouts today. For the first, I go back to the Twinings product mentioned in the first chapter of Part II, because its presence on my shelf poses a question. Its packaging states that it is a product of the UK and is blended and packed from imported ingredients by a UK company in the UK. And it was bought by me in the UK. So why does it have a 12-digit US generated number, rather than a 13-digit number resulting from an allocated company number by GS1 UK?

As we said earlier, fundamentally the question is irrelevant because the numbers are non-significant and can be generated in any country and read in any country. However, there is a particular bit of history which explains the presence of such numbers.

The earliest scanning and linked computer systems were developed in the US on the basis of the 12-digit UPC.

When systems emerged in Europe using the 13-digit EAN, they were equally capable of reading and storing the shorter 12-digit bar code and number. But the same was not true in the US, where scanning systems only continued to read 12-digit codes.
This meant that any manufacturer wishing to sell a product in the US had to include a 12-digit UPC on the packaging, and was faced by either using this on the same product in all countries or producing different packaging with a 13-digit EAN for the non-US market.

The problem was recognised at an early stage (and will be referred to at the end of Part III when looking at the development from two organisations to the single GS1 global organisation). But it is only relatively recently that many, though not all, US retailers have changed their systems to incorporate longer numbers and codes.

*****

The second subject is self-scanning. When I first conceived this book on retiring from my position as Chairman of EAN, I had a chapter devoted to the future of retailing which included this subject. It is a reflection of how much has changed in retailing, as well as how long it has taken me to finish this book, that the subject has been brought forward to be included in today's retail landscape. Almost every food supermarket I visit today has an area devoted to self-scanning. In most cases this involves taking the basket of goods to a self-scanning area and scanning each bar code in the same way that a checkout operator does.

In a smaller but growing number of instances, the store also has a hand-held scanning device carried around the store by the customer and used to read each bar code as goods are selected and placed in the shopping trolley. It is then this device which is 'read' rather than each item in the checkout area.

So what is there to say about self-scanning? In some ways there is nothing to say. As experts on the subject, you will appreciate that the operation of reading bar codes and retrieving prices is the same whether performed by a checkout operator, by you on a fixed scanner, or by you on a hand-held scanner. All of these are capable of reading the range of bar codes on products. Of course, if some form of weighing is necessary for, say, the pricing of loose fruit and vegetables, then this must be present on the fixed self-scanner, and cannot be done with a hand-held scanner without some weighing facility available within the store. And in the example of a reduced-to-clear price, the longer number system with the reduced price included allows those products to be self-scanned.

Having written that last paragraph, I visited a supermarket to ask what products I should note that could not be self-scanned and which required supervisor involvement to complete the transaction. The answer I got was that the product ranges which lead to most of the interventions by a supervisor are those which restricted for sale by age – generally alcohol but even items such as kitchen knives. There are also limitations on purchases by quantity etc on some pharmaceutical products, and particular items like gift cards and phone top-up cards could not be self-scanned in that particular chain.

But overall self-scanning in terms of this book is straightforward. Indeed, more than that, it is a natural development in the desire of retailers both to reduce the cost of the checkout operation and to speed the process at the checkout. However, it does raise an issue which goes back to the days of Ritty and Patterson – security.

Surely there is much more likelihood that a customer will avoid scanning an item which a checkout operator would automatically have scanned?

The supervisor I spoke to about self-scanning readily admitted that was the case. And all retailers are fully aware of this fact and accept that risk. However, that does not mean that they simply ignore it, and incidences are reduced by varying combinations of manual and technical scrutiny.

*****

To cover all products, all uses of bar code standards in all types of retail outlet in all countries of the world is impossible. In particular, this book has given less emphasis to developments and uses in the areas of the US and Canada, with their UPC history, than to other countries (and specifically the UK) with their EAN history. However, if it has given you a general understanding of what you find in the range of stores in which you shop, then it has achieved its objective.

But if there is much more that could have been said about the retail store environment, there is even more to be said about the retail industry in general. And so to Part III.

PART III – The New Retail World
(The Tip of the Iceberg)

16. It's All My Parents' Fault

Part II of this book was based on a customer perspective. In a store the customer sees the visible parts of the retail revolution – article numbers, bar codes and point of sale scanning. The effect of these is observable at the checkout with the benefits of much speedier processing than was possible previously and an itemised till receipt. But in truth this is only the tip of the iceberg.

So Part III will be based on the view of those involved in the retail industry – retailers, distributors, manufacturers, component and packaging suppliers etc. For they see what is happening below the surface, the bulk of the iceberg. The effect of the article number and bar code revolution is even more significant there, not simply in the way that it has transformed the management and operation of the 'under the surface' retail world, but in the benefits it has brought to customers in terms of price, choice and availability.

By chance, much of my life has been spent in that world. And by way of providing an introduction and overview of this unseen area, I will take you along my own learning curve. I must start with a couple who were responsible for my first lessons but who never lived to see either a bar code or the impact its arrival has had on the retail industry – my parents.

*****

My parents were shopkeepers. And if they had not been, this book might never have been written.

In 1954 they disturbed my pleasant suburban teenage London life when they bought a fish and chip shop in New Mills in north Derbyshire. For me it brought new friends (with strange accents and phrases), a new diet (with mushy peas, dandelion & burdock and chip butties), and Manchester United (in the days of the Busby

Babes). For my father it brought a move to an area close to his northern roots and family, as well as a new challenge.

This was neither gastronomic (he was a qualified chef, and his cooking skills were not necessary) nor business (for New Mills enjoyed its fish and chips in those days, even if the Regal Cafe has since become a Chinese takeaway). No, his challenge was to manage me.

It was not entirely my fault. Lovers of chips may not appreciate the hidden cost of what they eat. For every chip comes at a price – the potatoes must be peeled. We had an old bath in an old outhouse, which was always full of peeled potatoes. I remember it well. I am sure there must have been days when the snow was not on the ground, when the temperature was not below freezing and when I could feel my fingers – but I cannot remember them. My brothers insist that I was not the only one who exercised on the potato-peeling treadmill, which even with the help of a machine was physically and mentally demanding.

Though it was partly my fault. I could not be blamed for the fact that the peak Saturday sales at about 10.45pm when the pubs and Town Hall Hop turned out coincided with Match Of The Day. But I could be blamed for giving priority to watching United score their fourth goal (which I had seen at Old Trafford in the afternoon) rather than going down to the shop to serve a queue of customers that now stretched out of the shop door.

I had learnt my first retail lessons. Within the shop, my absence meant that chips were not cooked, and even the strong desire to walk home eating chips out of a newspaper did not prevent some lost customers and sales. Behind the shop, there was another world where bags of potatoes and peeled potatoes must always be available or chips would not be cooked at all. This most simple introduction to the importance of both customer service and stock management was soon followed by a broader understanding.

After a few years, my parents sold the business to take up the different challenges of owning a nearby greengrocery shop. Gone were the late evenings and coping with drunks.

In their place were early morning drives to Manchester Fruit and Vegetable Market and coping with customers who handled every apple in a box before buying.

In this relatively small community, customer service was paramount and 'retail was detail'. My parents knew every customer and every product. To lose a customer through poor service on only one occasion could result in a loss of hundreds of pounds each year, which was not easy to make up. To be out of stock of a product could be the trigger for such a customer to move to a competitor, yet to have too much stock was a major problem with perishable products. To price regularly purchased items too highly could be a further reason for losing a customer, yet to price too low would result in the death of a fragile business.

They had another fundamental decision to make. Was it better to travel to the wholesale market twenty miles away with the advantage of a wide choice of products and wholesalers but the disadvantage of travel costs and very long days? Or was it better to buy from a wholesaler who delivered to the shop, with the advantage of smaller quantities in terms of part-boxes, but the disadvantage of higher cost?

The detailed day-to-day questions they asked and decisions they took for their hundred or so greengrocery products in a single shop were essentially no different to those taken by the largest supermarket for their thousands of products in hundreds of stores. What products shall we stock (even if it was only a choice between varieties of apple) ; how shall we buy the products ; how much shall we buy (even if it was only one box or two) ; what price shall we charge ; what special offers shall we have ; when and by how much will we reduce the price to clear stock?

I would like to say that their attention to these details resulted in a successful business. Regrettably that was not the case and they sold the business after only a few years.

*****

Boots the Chemists was a bit different to my parents' shop. It had over a thousand shops spread all over the United Kingdom, of varying size, stocking thousands of products which were different from shop to shop. It had over a hundred buyers taking the decisions my parents took in the greengrocer's shop.
And it had several central warehouses supplying goods to the shops as well as a number of factories producing some of its Own Brand products. It was a bit different, and a bit more complicated.

My first job at Boots in 1966 was as a computer programmer writing part of a new system for the new IBM 360 computers. Although Boots had been using older EMIDEC computers for some years, the more powerful IBM machines provided the opportunity to do much more. The Merchandise Accounting system, as its name implies, kept records for the products bought and sold by Boots, most of which were supplied to the shops through central Boots warehouses. For the buyers, the figures for central warehouse stocks (though not individual shop stocks), historic supplies from the warehouse to shops, and any stock already on order from a supplier, allowed them to calculate future orders from suppliers for delivery to the warehouses. For the shops, receipts from the warehouse and their own stock counts allowed them to calculate their orders from a Boots warehouse. There were many sophistications within the system, yet for both buyers and shop managers it was fundamentally little better than my parents' basic system of counting stock, deducing what to order and noting those orders on a written or printed form. Indeed my parents almost certainly achieved higher levels of customer service and better stock management than Boots' staff achieved. But then the staff and systems at Boots did have to deal with millions of calculations and orders each week.

During the next few years new computer systems were designed to automate many of these calculations, reducing staff time involved and achieving more consistent and better customer service and stock management. The Wholesale Impact system replaced buyers' manual ordering of products from suppliers for delivery to warehouses.

The ASCOT (Automatic Stock Control Ordering Technique) system replaced shop staff manual ordering of products from warehouses for delivery to shops. One sceptic in the company suggested that ASCOT should have been named AINTREE because whereas Ascot was a flat horseracing course, Aintree was a steeplechase course renowned for the worst fences to be jumped and the most fallers. But both of these computer systems were based on the assumption that stock replacement was fundamentally a statistical exercise, albeit with the opportunity for human override where additional information was available. Therefore such automated systems would result in consistently better stock management.

For some products, even these automatic stock ordering systems were insufficient. If a customer came to a Boots dispensary with an NHS prescription, they did not expect to be told that an item was out of stock but would be available when the next delivery came a few days later. This was quite different to finding a particular toothpaste was not available and having the choice of an alternative size or brand. Customer service demanded that the item was obtained as quickly as possible, and certainly within 24 hours. Since only two central Boots warehouses supplied pharmaceutical products for the whole of the UK, they could only supply most shops with an overnight service at best.

So this led to a two-tier supply system ; the most urgent requirements were obtained by telephoned orders to local wholesalers for same-day delivery, and where the customer was happy to return the following day, those products were supplied the following day by Boots' central warehouses using an urgent order and delivery system. I was responsible for the development and introduction of a new system for these latter orders.

These were days before the introduction of the advanced technologies of personal computers and broadband telecommunication networks.

In the early 1970s, any replacement for a system of dictated telephone orders which were handwritten by warehouse staff had to be based on large mainframe computers or some local devices with limited 'intelligence' and erratic emerging electronic communications over telephone networks. The new system involved the dictation of orders by pharmacists in each branch to central regional branches, where they were typed on Teletype machines. The paper tape orders produced were then transmitted electronically in batches to the warehouses, where they were automatically printed for warehouse assembly and overnight delivery. The system did not need a product's description but was based on a five-character Boots alphanumeric code which identified each product uniquely. In today's world it sounds a very crude system. At the time, it represented a major step forward in using technology to increase service levels and reduce costs.

At the same time, I was also involved in the replacement of a manual system which involved the physical posting of printed buyers' orders sent to some of Boots' suppliers with a computer system which transmitted electronic orders, thus reducing the time for the receipt and delivery of orders. Once again the system was based on product codes, in this case a four-letter Boots item code.

But whilst all of these changes during the ten year period in which I worked with computer systems were important for the business, the real breakthrough was still a further ten years away. My involvement in that was a matter of luck because at this moment my learning curve became steeper when I changed jobs within the company.

*****

Timothy Whites was unlike my parents' shop in every respect, with only one exception - they both struggled to make any money. Following the Boots Company purchase of Timothy Whites and Taylors (a chemist and houseware chain) in 1968, rationalisation had led to a chain of over 1,000 Boots the Chemists shops and just under 200 Timothy Whites houseware shops. The former was highly profitable. The latter was not.

I became involved in 1977, initially as part of a small team which was given four of the Timothy Whites shops to run as experimental Cookshops, and later as Marketing Director for the chain. It would be nice to say that this move led to the transformation of the business. Such was not the case because the chain was closed down in the 1980s. However, for me personally, what I learned in this period was very significant.

I was frustrated. For the first time since my parents sold their greengrocer's shop, I thought about the management of sales and prices. My parents knew everything about the hundred or so products in their one shop.
They could stand in the centre of the shop and see them all. I knew very little about the thousands of products in what had become about 150 shops and could see none of them.

Yet the systems used by Boots and Timothy Whites in the late 1970s were among the most advanced found in UK retailers. As I have noted above, both chains were served by central warehouses and supported by computer systems for stock management in both stores and warehouses. But whereas the power of central mainframe computers had allowed these to be developed, no such computing power existed in stores. And so management information systems and decisions could not be based on <u>up-to-the-minute actual</u> product sales in stores, but only on <u>historic deduced sales</u> calculated from the product orders from stores to central warehouses allied to regular stock counts in stores. Of course these orders reflected what stores were selling (we hoped) but they were a long way from my parents' hands-on visual up-to-the-minute sales and stock information system.

Some retail chains did have better sales information. For example, some shoe chains put a punched card with details of the product in each box of shoes, and the card was then returned to Head Office when a sale was made. This information was used to replenish stocks but also meant that sales details were known. Some chains had short number 'velocity codes' on key products which could be entered into advanced cash registers to record each sale.

And some chains, including department stores, had introduced their own numbering systems for products which could be read by Optical Character Readers or key-entered at the till to give sales information. But in general the period around 1980 was characterised by limited and belated sales information systems.

In Timothy Whites, part of my job was to produce a programme of sales promotions. This involved working with buyers to select a number of products for special offers, determine the promotional price, and estimate how many extra would be sold and how much extra profit would be generated as a result. This activity has always been part of retailing, though for most chains it was much less significant as a proportion of their turnover at that time relative to today. Nevertheless, the effect of promoting a single key product at a special price which was lower than any competitor on the High Street was sometimes astonishing. I can still recall the dominance of unbeatable offers on products such as the Kenwood Chef in advertisements, shop windows and on the shop floor.

But all of this programme had to be based on inadequate historic and current sales information. As if this was not bad enough for the whole chain, the variation between stores in terms of normal sales and the uplift from price reductions was enormous. So I comforted myself that when store managers questioned the decisions, or even the sanity, of 'those at Head Office', in terms of the promotional sales programme, it could at least in part be attributed to inadequate information.

It led me to set up a simple system where store staff manually recorded each sale of a limited number of key products and reported the weekly sales to Head Office each Monday. By Wednesday buyers would have a full computer report of these sales, broken down to store level if required. It was a very crude system and was not wholly accurate, depending as it did on manually recording each sale as it was made.
But it was a small step forward. And the experience for me was significant in what was to follow.

*****

In 1982 I moved from Timothy Whites when I was appointed Director of Information Systems for Boots the Chemists. It was a return home, for I had started my career in Boots as a computer programmer and later as a computer systems manager. And sometimes in life you can be lucky enough to be in the right place at the right time. I was lucky in a number of ways.

In my new position, I replaced my predecessor on the Board of the UK EAN organisation. Although it had been seven years since EAN had been formed, the world of bar codes had not impinged on my thinking during that time. But now I was quickly convinced that it offered a major new opportunity for Boots. Although very few products sold by Boots had printed bar codes and the first point-of-sale scanning store had not been seen in the UK, I was sure that we were at a defining point in retail systems development. In particular, my earlier computer experience with Boots, allied to my retail experience with my parents and my marketing experience with Timothy Whites, led me to the conclusion that this new technology offered the chance to base decisions on up-to-date detailed store sales information for every product.

My second piece of luck was to work for Keith Ackroyd. For some within the Boots Group at the time, the Boots the Chemists chain was seen as a 'cash cow' – a business with shops in every town, high market shares, under attack from supermarkets and discounters, and with little opportunity for sales area or profit growth. Its profits were seen as best used in the expansion of other parts of the Group (pharmaceutical research and sales, and international retailing) rather than reinvestment in its own business. Keith had joined the company as a pharmacist and had managed shops of varying sizes, as well as being responsible first as the general manager of an area of shops and then as Area Director for a large part of the chain. Now as Managing Director, he was determined to refute the 'cash cow' argument and make a successful business even better. Furthermore he shared the belief that substantial investment in the new technology and the use of the information it produced was the way to achieve this.

To set that as an objective was a long way from its realisation and would not have been achieved as quickly as it was without a third piece of luck. The world of computing in 1984 was dominated by large central mainframe computers and, to a lesser extent, by smaller mini computers, each produced by different manufacturers. Personal computers were in their infancy with the first IBM PC launched in August 1981 and the first version with a standard internal hard drive in 1983. But I was convinced that the PC would become powerful enough to run our point-of-sale system and did not want to base a long term investment on mini computers. A visit to the United States in 1984 had seen five point-of-sale systems, but none met the requirements we had set. However, on our return, IBM asked us to meet them at a hotel near Heathrow airport, where they indicated that they not only shared our thoughts on the PC but were developing hardware and software based on the PC which could operate our largest stores.

The timing was perfect and in 1986 Boots became one of the first general merchandise retailers in the world to introduce scanning, controlled by a small PC in the back of a store. In the next few years not only was scanning extended to all stores (the tip of the iceberg) but the sales information provided transformed other parts of the retail business (the iceberg below the surface). None of that would have happened without my fourth piece of luck – a team of computer staff, retailers and trainers who were unbelievable in their professionalism, attention to detail, imagination and dedication. The names of those who have contributed to the transformation of this part of the retail world are never mentioned. But in listing Caird Biggar, Gil Duffy and Tony Kemmer, I pay tribute to many more in Boots and the wider retail industry who turned a technical possibility into an everyday reality.

To complete this personal journey, I must refer to a different system which I was responsible for introducing at this time, which links back to my earlier experience with the overnight receipt of orders for pharmaceutical products. Boots had many shops with limited space in prime locations whose opportunity for growth required a larger selling area.

One way of achieving that was to find a way of reducing the space taken for back-store stock and converting such space into an enlarged sales area. But in order to do this, the existing system of weekly (or less frequent) orders and deliveries for outer packs of products stored there needed to change to daily deliveries of single products which could be put directly into the sales area. This in turn led to the establishment of small regional warehouses (referred to as Common Stock Rooms) for the best selling products which could accept and deliver orders every day. Again we were fortunate that the development of hand-held devices to record stocks, telecommunication to transmit from these devices and personal computers to process the figures and produce orders allowed a system which would not have been possible a few years earlier.

*****

My experiences above have said nothing about the retail check-out. They have all been about the hidden aspects of retailing which will be considered in this part of the book. The rest of Part III will show how the retail industry has used article numbers, bar coding, scanning and sales information to transform stock management, logistics, sales promotion, space management, administration and customer service. The starting point will be stock management. At the age of fifteen, the subject meant keeping a bath full of peeled potatoes. I know now that there is a bit more to it than that.

17. They're Out of Stock Again

It is market day today in my home town of Wirksworth. And we have decided to make a stew. This seems a sensible choice on a day when the north wind is whipping through the town and snow flurries are forecast. But shopping starts badly. Our visit to the market square late in the day reveals that the stallholder selling fruit and vegetables must also have felt the wind and seen the forecast – she has decided not to turn up. But we move on to one of our local shops selling vegetables. However, the swedes and parsnips we intended to use are not available. We assume that shop sales have been unexpectedly high because of the absence of the market stall. It is frustrating and our plans must be modified. Gone are the days when consumers bought what happened to be seasonally available and on the shelves. We expect anything we want to buy to be available whenever we want it. Putting aside for the moment whether this is a sustainable or even a desirable development, it is a tribute to the retail industry that such expectations are almost always met.

However, it is not always easy for retailers. And my elder daughter does not help them. When her favourite brand of apple juice is promoted in her local supermarket as '2 for the price of 1', she clears the shelf of all stock. I can almost hear the next potential customer muttering as they survey the empty shelf, 'They're out of stock again'.

So how do 21st century retailers try to ensure that they satisfy all their customers, for all of their products, all of the time? And what difference has the new world of article numbers, bar codes and scanning made to product availability and customer service?

Today, each item sold is recorded as its bar code is scanned at the checkout. The power of computers and transmission speeds mean that every sale can theoretically be known by the store manager, the back store staff, the central warehouse, the Head Office buyer, the manufacturer, the ingredients suppliers, and the product packaging supplier almost instantly.

Of course, that does not actually happen. But each sale can, and often does, adjust the figures on a retailer's computer files to reduce the theoretical stock of that product still available and increase the total sales figure for the particular product. The implications of this for stock management systems are profound.

Before the introduction of point-of-sale systems, the starting point of such systems, whether paper-based or computer-based, was the stock count. For many retailers, systems were based on an order-up-to level (OUTL), which meant that the stock to be ordered was simply the OUTL less the current stock level (adjusted if the order had to be placed as a multiple of an outer pack size). Periodically this OUTL was increased or decreased according to the rate of sale. For other retailers, a calculation of how many items had been sold was made by taking the previous stock count, adding any quantities of stock delivered and subtracting the current stock count. The deduced sales figure for the period meant that the rate of sale per day or per week could be determined. This rate of sale could be matched against the current stock figure leading to a stock order reflecting the changing rate of sale rather than the fixed OUTL.

After the introduction of point-of-sale systems, the starting point of such systems can be the rate of sale because every sale is automatically recorded. And since the available stock figure is reduced by each sale, the key figures of current stock and rate of sale are both automatically available, allowing an order to be directly calculated. At this stage it is sufficient to note that these up-to-the-minute figures allow major retail chains, with hundreds of stores, tens of thousands of products in each store, and many millions of stock replenishment calculations to be made each week, to achieve very high levels of stock availability.

That's the theory. But is that what is actually happening behind the scenes in the stores you visit? To answer that, I will not start with the largest of the supermarket chains but work upwards from the smallest of my own local shops. After all, Kens and Dips here in Wirksworth both have scanning systems in their small family grocery shops. So what difference have those systems made to their management of stock? Oh dear, the answer is *none*.

That is not quite the answer I hoped for when planning a chapter to highlight the stock management revolution. But seemingly unhelpful answers may actually prove useful in understanding an issue.

Both of these shops have scanning systems on their counters, which have allowed them to eliminate individual pricing of some of their products and to serve their customers more quickly. In doing so, their point of sale systems are recording every sale made for each product. And each of their systems has an optional facility to use that sales information for stock reordering. So why are they not using those systems?

Both of them told me that their stock ordering systems actually continue to be based on walking round their shops, looking at the stock on their shelves, and writing their orders in a notepad. In other words, the same system they have used for years and that my father used half a century ago. One of them gave me this information rather sheepishly, as though I and others would regard it as an indication of being old-fashioned. But he then went on to explain that in a small shop the hands-on detailed knowledge of all his products, sales and reasons for daily variations meant that he had much more confidence in his own calculations than those of an automated system. Furthermore he used a number of different suppliers for his goods, and his computer system was independent of all of them. In other words, stock management based on stock counting remains for them both the most cost-effective way of conducting their businesses.

When I move on to a somewhat larger third grocery store in the town, I find a different picture. The Spar shop is independently owned but is part of a franchise operation using a common point of sale system, common back office software and with goods supplied from Spar central warehouses. As such it is somewhat easier for it to use sales information as the basis for stock ordering, and it does so. The in-store computer system indicates to the manager the theoretical stock of each item and produces recommended orders, which can be authorised or modified.

These orders are transmitted via the Spar system for supply and because subsequent deliveries are also automatically recorded by the system, stock figures are updated. So, if all these stock movements are built into the computer system and its calculations, is there ever any need to count the actual stock on the shelf? Oh that life was so simple.

But, as one of the staff told me, '*We still need to count stock on the shelf so that the computer's theoretical stock figure can be corrected for breakages and items that leave the shop without being paid for!*' So her words remind us that the problems of Ritty and Patterson and goods which '*leave the shop without being paid for*' are alive in the 21st century and impact even these more sophisticated stock management systems.

Whilst the three examples I have mentioned are all general merchandise stores, the use of scanning and linked stock management systems is not confined to such stores. To illustrate this, I note a fourth shop in Wirksworth, which installed the first scanning store in the town.

Nick Payne is a pharmacist. In fact he is the third generation of his family to be a pharmacist. When his grandfather bought the chemist shop in Wirksworth in 1933, it had already been in existence on that site for 177 years. Nick qualified as a pharmacist to dispense medicines in the shop in the late 1980s. And in 1990, 24-year old Nick persuaded his father to borrow £8,000, a considerable sum for such a business at the time, to buy the first Electronic Point of Sale system in Wirksworth. Yet Paynes the Chemists was a small shop run as a family business with no significant problems of customer queuing, sweethearting or hands-in-the-till. This raises the question of why this business invested in such a system in these very early EPOS days. And the answer is that, because stock management was so important and so administratively time consuming, the investment was justified simply to improve that aspect of the business.

Of course this is only a sample of four shops and is not intended to be seen as indicative of all such retailers. However it does emphasise that the presence of technology and the option of alternative ways of working does not automatically mean that they offer the best solution in all circumstances although it can be beneficial in small as well as large stores.

In moving from these shops to the major retail chains you use, it can be assumed that all of them, whether food stores or non-food stores, will make use of automated stock management systems. Because this involves vast numbers of stock movements, calculations and decisions, the subject can seem overwhelming. So it is necessary to simplify it, because after all 'stock management' comes down to individual products in a specific store (stock-keeping units or SKUs). And to understand the stock management process, linking it back to some of the understanding of article numbers from Part II, it will now be reduced to my purchase of one particular product in a particular store of a particular retail chain.

*****

It was not on my shopping list and I had no intention of buying it. And I would like to think that I am an intelligent shopper, well aware of the cunning actions of marketing managers tempting me to buy what I do not necessarily want or need. But I could not resist this special offer. There in a prominent position was an offer ' *buy 4 get 4 free'* for Muller yogurts with no restrictions to the mix of different flavours bought.

But this is not the moment to write about the psychology behind my action.

It is not even time to get into a debate on why the retail industry in the UK has changed the spelling of *yoghurt* to the Americanised *yogurt*. Instead we will focus on the stock management of Muller yogurts in this store.

The different flavoured products included :

Muller Light Yogurt - Cherry	402 550015737 1
Muller Light Yogurt - Strawberry	402 550015734 0
Muller Light Yogurt - Mandarin	402 550015739 5
Muller Light Yogurt – Rhubarb	402 550015745 6
Muller Light Yogurt – Vanilla	402 550015747 0

In passing we should note, as shown in Part II, that each flavour when sold separately has a different article number. All of the products start with the digits 402, indicating that GS1 Germany gave Muller its company number, although the yogurts came from Muller's factory in Market Drayton in the UK. The product has a limited shelf life, with a best before date of 14 days from when I purchased it, and is sold from a chiller with a limited capacity in the store.

If this 'buy 4 get 4 free' offer had consisted of a pack of 8 products, only the unique outer bar code would need to be scanned. However, for my purchase, the scanning system must record the purchases of the varying quantities of the five different flavours I have bought, including three of one flavour and two of another. The data will also be used to determine any further orders and deliveries to the store. The relatively limited sales of the Muller range within the store mean that these products will have been supplied from the retailer's central warehouse, which will in turn have received much larger quantities of stock direct from the Muller factory. The sales of these particular products will be affected by the promotion, though there may be differences between the uplift to sales of different flavours. It is possible that all of the stock of some or all flavours will be sold in a short time, so the use of immediate sales data to speedily replenish stock automatically is critical if sales are not to be lost. The frequency with which the sales and stock holding of these particular products are reviewed will depend upon the stock management system in use and the number of deliveries made from the central warehouse to the store.

But central to all this activity is an accurate figure of sales automatically consolidated from each individual purchase made across all the checkouts in the store. Contrast that up-to-the-minute information with the pre-EPOS system, when sales could only be deduced periodically when store staff counted the stock left of the product in the different locations it was found in the store.

Of course, the use of sales data does not mean that all stock counting is avoided. As the Spar example earlier noted, regular stock counts are still necessary to correct the theoretical stock held by the computer and to bring it in line with the quantity of the product found on the shelf. We have already seen that breakages and theft are two reasons for any divergence. But a third factor is any error made in the scanning operation. Any mistakes made at the checkout will affect the sales figure for the product and its remaining theoretical stock. Consider how such errors can easily occur for our related products.

Suppose that at the checkout, I tell the operator that I have 8 yogurts and, whether it appears that all are of the same flavour or in order to save time, he or she scans the bar code on one of the products and enters that 8 have been sold. Unfortunately, the result of this is that the computer system records that 8 of one flavour have been sold but none of the other flavours. This explains why from the outset EAN and UCC laid down that every item, including flavour variations, must have its own unique article number and why each separate product must be scanned at the checkout. The stock management system is dependent on this information.

All of these calculations lead to orders and deliveries to the store. But although some products may be moved directly onto the sales floor from an incoming delivery, many others will have further stock available in the back of the store. So, if stock ordering systems based on immediate sales have impacted deliveries to stores, how has the introduction of scanning impacted the movement of goods from the back store to the sales area?

And to illustrate the potential change that has made, I return to my daughter, who has just cleared the shelf of a particular brand of apple juice.

Suppose that the computer in the back store area knows how many of this product can fit onto the shelf, are actually on the shelf, and are in the back store. And then suppose that it is given every sale of a product at the moment it is scanned. Surely that means it could alert back store staff to the fact that my daughter's action has left no products on the shelf, but that there are products in the back store and even where to find them. In this way, the shelf could be refilled as quickly as possible and no sale would be lost. There are only two problems with this. Firstly, there is a time interval between a customer taking a product off the shelf and having it scanned at the checkout. Secondly, as we have noted earlier, the computer stock figure may not be accurate because of other factors.

The first of these is easily addressed. If the shelf space is sufficient for, say, three outer packs of a product, the back store computer can trigger replenishment when the sales of two packs have been made and before the shelf is bare. The second can only be dealt with by regular stock counts (also using the bar code and hand-held scanning devices), when the actual shelf quantity can be corrected (and, incidentally, an accurate figure can be given for the amount of theft of each product in the store).

Earlier in this chapter we saw that even though a technical solution was possible, a retailer might decide that the manual alternative was preferable. And so in this case some retailers may choose to manage the replenishment of shelf stock from the back-store area in the way we have described. But others will feel that it is more efficient to look at the shelves of a product area, to record the gaps that need to be filled (possibly using a scanning device) and to replenish from that list.

*****

This chapter has introduced the subject of stock management. Its limited scope has been to show that the accurate scanning of bar codes at the checkout is the critical starting point for the calculation of replenishment orders. The next chapter will go into more detail in terms of the change which article numbers have made to the physical movement of products resulting from those orders.

And the following chapter will consider equally significant changes to the communication of orders and other product information between retailers and suppliers.

18. Making Connections

So far, so good – I hope. Every different product has its own article number. The scanning of the bar code associated with that number is used not only on the surface at the checkout to speed the sale to a customer but also under the water as raw data to trigger a re-ordering process. And when that order is made, its communication can be made presumably in terms of its unique article number. Unfortunately life is not quite that simple and we must now introduce more numbers and more bar codes.

The reason can be seen clearly on some of the occasions when I visit Ken's small grocery shop. The floor contains boxes of items which have recently been delivered and need to be unpacked and the products put onto the shelves or into chiller and freezer cabinets. The movement of goods between a wholesaler or retail warehouse and a store is generally not in terms of the scanned 'consumer unit' but in terms of a larger 'despatch outer'. And going back one stage further in the supply chain, the movement between a manufacturer and a wholesaler or retail warehouse is generally in terms of a pallet containing a quantity of the despatch outers of a product. For those involved in dealing with these packages and pallets, the consumer unit is relatively unimportant. They need to identify either this larger package which is the despatch outer or the pallet of goods or both.

Therefore, whereas scanning systems at the checkout only require details of each consumer product, ordering systems must know not only about the single product, but also about the larger packs in which it can be ordered and delivered. Furthermore there must be some link in computer systems between the two if sales and stock counts of the consumer unit are to be converted into orders for the despatch outer of the product.

In Part II we saw that each product scanned at the checkout has its own unique Global Trade Item Number GTIN (of 8, 12 or 13 digits) and associated bar code (eg Baxter's Red Lentil & Vegetable Soup 415g).

But how is the despatch outer of, say, 12 cans of Baxter's Red Lentil & Vegetable Soup 415g, identified? And what link can be seen between the GTIN given to a single can and the GTIN for a box of 12 cans?

For retailers and manufacturers in the 1970s, the subject of identifying despatch outers was as important as that of consumer units. As we saw in Part I, whereas the history of bar coding concerned improving the checkout operation in stores, the history of article numbering, particularly in France and Germany, concerned this ordering process between retailers, wholesalers and manufacturers.

And here there is a story to tell of major differences of opinion between the members of the European organisation in the 1970s in their search for a solution to the numbering and coding of despatch outers. It may seem strange to you that a group of intelligent individuals should become so passionate and spend so much of their lives on the issue of numbers for despatch outers. To understand it you need to appreciate the mindset of those involved in computer system design.

As I noted earlier, part of my experience was in that world. The appeal of computerising and improving manual processes was the skill required to reduce them to the best set of logical instructions which would work in every circumstance for years to come. For whereas manual errors can be quickly traced and corrected, computer errors can be more complex to resolve. To many that would be utterly boring but to me it was stimulating. Though I can still recall almost losing my job when making an error in one out of thousands of lines of computer code which had taken me six months to write and test. The Chief Accountant told me that failure to correctly update the field *'Output to Casemakers Plus'* in a relatively little used process was almost a hanging offence!

So for these systems specialists, with this perfectionist mentality, the desire to produce the best international common standards which would be in place for decades to come was overwhelming. Their different ideas meant that the debate was protracted.

185

You can judge for yourself whether they came to the best conclusion.

*****

Having read this far, you will know that the EAN group of countries had already reached agreement on the standards for products sold to consumers – a unique 8 or 13-digit number for each product and an associated bar code, which could be used by any company in any country. So surely agreement for packages sold by manufacturers would be relatively straightforward? A simple solution would have been to say that every case or despatch outer or pallet of particular products should be dealt with in exactly the same way. Each one was unique and should therefore have its own 13-digit number and bar code. The number of the individual product would be different to the number of the despatch outer, but a table of numbers could be held on computers which linked the two numbers together. In the United States, with their 12-digit numbers, they came to this conclusion. In Europe, Germany was the leading country, but not the only one, pressing for this solution too.

The advantages were clear. Firstly, there would be a single numbering system and bar code solution. This would apply to every 'article', even if that article was a pack of a product. Secondly, for those retailers and wholesalers who sold both single consumer units and larger despatch outers of products, a single computer file of numbers and a single bar code reading device could be used for everything they sold.

The disadvantages were also clear. Most obviously, if a consumer unit was given one number and a despatch unit of the same products was given a different number, there would need to be some way of linking the numbers to each other for ordering purposes. This led some countries to propose an alternative solution, where the Despatch Outer should be identified by having the product's 13-digit consumer unit number followed by a multiplier suffix indicating the number of consumer units contained in the outer.

186

If there was disagreement about what number system to use, life was made even more complicated by the fact that there was also disagreement about what bar code system to use. To those involved in the industry, the bar code which had been designed specifically for small consumer units was far less suitable for the corrugated packaging of larger despatch outers.

This resulted in many countries insisting that a different form of bar code was necessary for despatch outers with different reading devices for warehouse and back store use.

And so, from an early stage two different number solutions and two different bar code solutions were being considered. The difficulties of finding a compromise solution, and one which fitted into similar discussions taking place in the United States, are apparent from the fact that agreement was not reached for five years. It was not that the problems were ignored and the issue put on the back-burner. The minutes of every meeting during this five-year period report on the progress, or rather lack of progress, in finding any solution.

But the world does not stand still, and during this period some countries could not wait for a solution to emerge. In these countries, retailers, wholesalers and manufacturers wanted an immediate solution – though not the same solution in each country – and would not all wait for a compromise to be reached. So, for example, Germany announced that they were proceeding with 13-digit numbers only, whilst Sweden indicated that they planned to go ahead with 13-digits plus an add-on quantity. But many countries did wait for a definitive position to emerge.

The next few pages could be filled with the factors and discussions which occupied meeting after meeting on this subject. But they would only serve to emphasise the ongoing differences of opinion. So what was the final outcome?

The absence of any single agreed solution finally meant that two alternative numbering options were defined and allowed.

The first accepted a unique 13-digit for every traded item, whether it was a product sold to a consumer or a Despatch Outer. There would be no direct link between the numbers given to the consumer unit and the despatch outer of the same product. The second option did not accept the original add-on quantity proposal, but did create a new 14-digit number consisting of the first 12 digits of the product number of the consumer unit (the check digit excluded) preceded by a single digit 'logistical variant' and followed by a newly calculated check digit for the longer number. Therefore, in this option there was a direct link between the number given to the single item and the despatch outer's item number.

Example :

Tone 50 g Dark Chocolate € 1   7614500233115
Tone 50 g Despatch Outer of 30 packs   27614500233119

As well as allowing two number options, the compromise allowed two bar code options. The first simply accepted the existing bar code for the 13-digit number.
The second introduced a different type of bar code (though still consisting of a set of lines of varying thickness) known as ITF-14 (Interleaved Two of Five) which would be used 'where printing conditions require a less demanding symbology'.

Two numbering options and two bar code options. In reaching this compromise, there was a recognition that it was far from ideal and could lead to future problems.

To minimise these problems, it was stressed that each country should only use one option and that companies in one country should not enforce their country's option on a company from another country which used the other option. However, it is interesting to note that the meeting minute which states that position also makes clear that it will be difficult to enforce that by noting that the members are *'aware of the limited power to sanction any abuses'*.

In other words, it was recognised that a particular retailer or manufacturer in one country might only agree to trade with one in another country if products were marked in the way that company wanted. And if and when that happened, some companies would be forced to allow for two number systems and/or two bar code marking and reading systems.

And was this rather messy agreement the best conclusion? It was argued at the time that it was essential to reach some agreement. After all, there had been five years with a lack of any common standard or agreement and it was inevitable that, without some statement, an uncontrolled position would result in no common global standards leading to future problems in international trade. And this was the best compromise available. There is no simple answer. It only illustrates the difficulty of reaching any uniform standards across an industry and across countries. However, without an agreement, extra costs and complications would result for years to come. Once established, such decisions are not easily changed.

I was not one of those involved in this decision and therefore have the luxury of hindsight in judging the actions of many of my later colleagues. It would be easy to criticise the length of time they took and the absence of any single conclusion at the end of that time.

And it could be argued that the consistent German position, which argued that the increasing power and speed of computers meant that their unique 13-digit solution requiring tables to link numbers would be simple and inexpensive in the medium term, was vindicated. So in that sense, perhaps insufficient weight was given to longer-term considerations. But businesses need practical solutions today and cannot always wait for a better long-term answer. And, whatever its faults, the compromise of optional number and bar code solutions was implemented and is still in use today.

*****

But if that is the historical background, what actually happens in the ordering and delivery process, regardless of which system is used? To see that we must return to my purchase of several flavours of Muller yogurts. As a result of my transaction, and those of other customers, the store ordering system will have determined that further stock is required in the store.
These orders may be placed automatically but may be subject to review and amendment by the local store management. But they will not be placed using the GTIN for the individual products.

The computer store ordering system will have converted a requirement for, say, 144 Raspberry & Cranberry yogurt into an order for 6 despatch outers each of 24 items. And it will have linked the GTIN for the individual item to the different GTIN for the outer pack of 24 products, placing the order in terms of that despatch outer's number. When the order is processed by the retailer's central warehouse and delivered to the store, it is the bar code for that despatch outer which will be scanned.

A similar process occurs for orders placed by the retailer's central warehouse with the manufacturer. The warehouse will have sent out despatch outers to a number of its stores and will need to replenish these from the manufacturer. Once again a translation must take place because Muller will not deal in despatch outers but in manufacturer's outers. This even larger pack of goods (possibly a pallet of goods) will consist of a number of despatch outers.

So another GTIN is needed for this larger pack. In other words, working down the chain from Muller, the manufacturer's pack must be broken down into despatch units by the retail warehouse and sold as individual products by the retail store. And the communication of this physical process will be in terms of three different GTINs which relate to the different sizes of the pack involved.

I could take this process much further. After all, Muller needs to order the components for its factory from other suppliers and each of these components can have its own GTIN and bar code, allowing a common communication language for ordering and delivery. But I think the logistical chain will be clear to you.

Equally, I am sure you will now appreciate the significance of store scanning systems to the ordering systems all the way back to component suppliers. The immediacy of actual sales figures means that the whole supply chain can be warned of any changes in demand, gear up their deliveries to adjust to these changes, and provide better customer service levels on the shelf than were ever possible historically. I gave the last chapter the title, 'They're Out of Stock Again'. In practise that is totally unjust. The remarkable fact is that modern retailing has led us to expect that, they will never be out of stock.

*****

Before completing this chapter on stock management, I need to expand (or complicate) matters by opening the door to another part of this behind-the-scenes retail world. All of the physical activity in the movement of stock has referred to the requirements for trade item numbers and associated bar codes. But is it possible that the movement of goods manufacturer to store receipt would benefit from other information shown on packages in the form of bar codes which could be read electronically?

The initial answer from the EAN and UCC organisations was *NO*. The logic was simple.

If the GTIN identified any traded item uniquely and all of the associated information about that product was held on computer files, the number was the key to accessing that information.

In other words, don't complicate things by printing other information on packaging because it could be retrieved by use of the GTIN from computer files. Why should any of that information also be printed physically?

But in practice, access to that information from computer files was not always available wherever it was needed. For example, information such as production batch numbers, weights and sell-by dates were already being printed on despatch outers and cartons. So why not find a way in which these important items of information could also be translated into bar codes so that they could be read and allow checking goods during movement to take place? A system was required which would allow a set of key data to be printed in bar code format in addition to the basic item or package GTIN.

This led to the development of what became known as Application Identifiers (AI).

A limited set of 10 applications was introduced in 1989. Each consisted of a two digit identifier followed by a variable length field which could be numeric and/or alphabetic. For example, the manufacturing batch number was given the AI 10 and the length of the actual batch number could be anything up to 20 alphanumeric characters, reflecting the different formats used by manufacturers. This AI could be represented in bar code format, though the pattern of bars followed a different format to that we saw in Part II. The format was known as EAN/UCC-128 (now GS1-128) and, though its detailed structure will not be given here, reading the lines of it can be found on many websites. With this new system, not only could a single AI be printed on a pack with a bar code but several different AIs could be run together and printed one after the other in one bar code.

Since their introduction, the number of Application Identifiers has grown enormously, reflecting the importance of their introduction and the growth in use of automatic reading equipment throughout the logistics process. An AI can now start with a number up to four digits and the following data may be of fixed or variable length with some AIs extending to 30 characters.

AI 00 has been developed to identify uniquely any item transported within the logistics chain (say a pallet of goods containing one or more products) and is referred to as the Serial Shipping Container Code (SSCC). This allows it to be tracked through warehouses and deliveries. Other AIs can give details of various dates (production, packaging, expiration), physical attributes (weight, dimensions, volume), delivery addresses, order numbers etc.

Of these AIs 410 to 414 are of particular interest. All are followed by what is referred to as the Global Location Number. Although the original idea for a 13-digit number was to identify articles uniquely, as early as 1980 it had been accepted that the same concept could be used to identify locations uniquely. And so the 13-digit GLN can be used to indicate a despatch address, a delivery address, an invoice address or simply a company legal entity. Just like the GTIN, the GLN is unintelligent and non-significant, but the unique number can give access to computer files offering more information associated with this number.

The ability to scan bar codes giving all this additional information means, for example, that pallets and packages can be automatically routed, that any packs which have passed a sell-by date can be identified, any batches with a manufacturing problem can be extracted etc. It has transformed the physical process of moving goods efficiently and accurately.

Unlike the products passing the store scanner, these items with their associated alphanumeric identities, bar codes and scanners are not visible to the store customer. They are below the waterline. So, despite their significance to stock management and customer stock availability, I have not detailed their structures and they have occupied only a part of only one chapter of this book.

The same can be said for the next linked subject. Because, whilst this chapter has been devoted to a physical set of movements and information, the next looks at the electronic world which relates to it.

19. Computer Shall Speak Peace Unto Computer

For my granddaughter it was nothing short of a disaster. Her best friend was going to live in Spain for three months. The two ten-year olds live only 100 yards apart and spend so much time together that the separation would last for what seemed an eternity. But they are children living in 2013 and life is not what it was as these examples show :

- In 1953 when I was a teenager any communication would have been by letter sent by sea and land transport and taking several days.
- In 1983 when my eldest son was a teenager the link would have been either by letter or using our fixed-line telephone. The first IBM PC with a hard disk was launched in that year.
- In 2003 my youngest daughter could add the option of long evening email conversations using our desktop computer. The Berry household had not bought our first mobile phone and Skype was only about to be launched.
- But it 2013 my granddaughter has a range of choices, from email to Skype and from Skype to Ipods. She can not only communicate but also see her friend every day. Life is bearable after all.

I have listed these changes because anything written on the subject of electronic communication in the retail industry is likely to provoke the reaction, 'So what! How else do you expect to communicate?' Therefore some historical prospective is given here of times before all today's options were available.

The basis of the subject really goes back to a key idea from Chapter 3 of this book. Nearly 50 years ago, Karl-Heinz Severing had said :

'The food trade has just started to introduce electronic data processing. Perhaps 1000 companies could use such systems but fewer than 50 have been installed so far. With EDP, retailers and wholesalers are just a step away from integrated data processing with their suppliers. For example, it is possible to process automatically all incoming invoices without manual processes. It is conceivable that the supplier could provide a 'punched card' instead of a printed invoice which would contain all the invoice details and could be entered directly into the computer.'

The language then was of electronic data processing, of punched cards and later of a retail industry number system. The language now is of computers, electronic data interchange (EDI) and a global multi-industry system of numbers and standards. The realisation then was BAN-L in Germany with its list of numbers printed in catalogues and on orders and invoices. The reality now is EDI throughout the world.

To understand this reality, we need to go back to the problems which Severing was trying to address, but with the added advantage of looking at them with the knowledge of what today's technology makes possible. He wanted to get rid of paper orders and invoices, and some of the manual processes required to deal with them. We assume that almost any transfer of information in the retail industry can be achieved in a paperless way.

Today, the development of technology means that numbers and messages are transmitted almost entirely from computer to computer using EDI. The three elements of EDI are the transfer of structured data from one computer system to another, the use of agreed message standards, and the absence of any human intervention in the process. The aim of the message standards which drive these communications is to provide unique, unambiguous messages understood and agreed by the sender and the receiver. Of course they result in a much speedier and less expensive form of communication.

But importantly, they are also designed to take away all the differences, mistranslations and misunderstandings which might lead to errors, arguments and battles either over goods that are ordered, delivered and charged or other information needed for trade.

In the 1960s, the British satirical revue *Beyond The Fringe* hit the London stage. One of the sketches featured the British Prime Minister of the day, Harold MacMillan. Its fictional account of his meeting with the German Foreign Minister reported him saying of it, *'We spoke many frank words – in our respective languages. Though precious little came of that in the way of understanding'.*

In 1927 the BBC introduced a coat of arms with the motto, *'Nation Shall Speak Peace Unto Nation'.* This phrase was designed to indicate the values of the organisation and was a feature I remember from my childhood when crouched in front of a small black and white television set waiting for a programme to start. In understanding part of what is happening with EDI behind the scenes in the retail industry, we might coin the phrase *'Computer Shall Speak Peace Unto Computer'.*

*****

So what information needs to be transferred between trading partners? Unfortunately the list is rather long and, since it is not the aim of this book to be comprehensive, we will limit this chapter to some of the most common elements. Obviously, the starting point is the specific product which is being bought or sold, invoiced or paid for. We have already seen that individual products and larger packs of products, including despatch outers, are identified by unique article numbers. This means that the computer files of both the seller (manufacturer) and buyer (retailer) of any product contain this number. But the number needs to be associated with a description of the product. In fact several descriptions of the product will be required because the length and detail needed on an order or invoice will not necessarily be the same as that required on packaging, in the supplier's warehouse, on the retailer's shelf-edge label or on the customer's till receipt.

Other characteristics of the product will be required such as the physical dimensions, the weight and its ingredients. For any particular packs of a product it will be necessary to know the batch number of its manufacture, date of production, possibly a sell-by date or best-before date. And for any order and delivery, despatch, delivery and payment addresses will be required. Furthermore details of the cost and selling price of the item, discounts and payment terms will be required. After an order has been delivered, messages will still be required to indicate any differences between the quantities ordered and those delivered, and details of any payments made. The list goes on and on.

For the manufacturer, these details will need to be sent to many retailers and wholesalers, probably in many countries. For the retailer, the details will need to be obtained from many manufacturers, also from many countries. So the potential to transform this vast complex of paper messages into electronic format sent from computer to computer is enormous. And more than that, the potential to pool some information in order to reduce the number (though not the detail) of those messages exists too. What does it require to achieve all of that?

And there is the answer. Electronic communication requires a common language and a common understanding. If this mass of information is to be sent from computer to computer without human intervention, it implies that all the large number of different messages must be produced in a unique standard format if they are to be received and understood unambiguously. And just as you have seen the standards which were developed for article numbers to ensure their uniqueness, so there have been standards developed for all of these data elements to allow their communication.

The leading UN/EDIFACT standard comprises a set of internationally agreed standards, directories and guidelines for the electronic interchange of data. The first two standard messages were produced in 1988. Perhaps not surprisingly, they were for orders and invoices.

The extensively used GS1-EANCOM is a subset of these standards and was developed to provide a somewhat simpler set of messages which contained all necessary information but excluded elements relating to rather specific businesses. The messages allowed the physical movement of goods to be integrated with related information which was sent by electronic means. All of these messages (for trading, transportation and finance) are equivalent to traditional paper business documents and cover the elements required to complete a trade transaction. The set of 49 different messages is currently used by over 100,000 companies in over 150 countries. Other message standards exist, as we shall see shortly.

The messages remain necessarily very detailed so all the elements and format of any single message such as PROINQ (Product Enquiry), IFTSTA (Transport Status), RETINS (Instructions for Returns), and SLSRPT (Sales Data Report) are not given here.

In addition to my experience with the Boots retail businesses, I was briefly involved in the 1970s with implementing a new computerised despatch and invoicing system for Boots International, which exported pharmaceutical and healthcare products. This made me aware of how much more information was required for international trade documentation and how different this information could be for different countries. These complexities had to be accommodated within the new EDI standards and can be seen within, for example, the UN/EDIFACT for ORDERS.

The definition of this is simple enough as '*a message specifying details for goods or services ordered under conditions agreed between the buyer and the seller*'. But its scope is apparent when it goes on to say that, '*The Purchase order message may be used for both national and international applications. It is based on universal practice related to administration, commerce and transport, and is not dependent on the type of business or industry*'. As a result, although the standard defines a header section which will ensure that a receiving computer system can identify a particular message as an Order and refer to this order uniquely, the devil is in the detail.

To cater for the vast range of information that may now be required within an order, the standard goes on to set out how these details should be encoded in 60 separate segments which are divided into 170 message elements.

So perhaps it is not surprising that early attempts to establish communication standards focussed on the somewhat simpler transactions taking place within national boundaries. Even by the mid-1980s, only a relatively small percentage of orders involved international trade, so the focus on automating the greater number of national orders was both desirable and inevitable. As a result, a number of national communication standards emerged, each with their inevitable acronyms, such as DAKOM in Sweden, ICOM in Belgium, GENCOD in France, SEDAS in Germany and TRADACOMS in the UK.

The system within the UK was launched in 1982 and was extensively used by the time further development activities were switched in the mid-1990s to the global EANCOM standards. However, it is an indication of the importance of such standards to the day-to-day running of businesses that even now a substantial number of companies continue to use these TRADACOMS standards. As we saw in the case of article numbers, it is a lengthy process to determine, establish and expand any global standard. And it is equally the case that such standards as TRADACOMS, once established, are time-consuming and difficult to replace. They represented a breakthrough when they were introduced, have led to new ways of doing business, and for many companies their replacement is not a priority. And so TRADACOMS, GENCOD, SEDAS, JCA in Japan, CNAB in Brazil and the international standard ANSI/X12 in North America continue to be widely used.

But the world does change. New technologies emerge, new ways of doing business follow, new requirements for communication between trading partners are needed and new standards are introduced to accommodate this new world. So, in practical terms, how do companies now communicate with these standards? Of course there is no single common answer.

But to understand the options and opportunities that are now available and in use, a number of other developments under the auspices of GS1 must now be introduced.

*****

I start with the Global Data Synchronisation Network (GDSN). The idea behind this is a very simple one. Is it possible to have one simple storage area where companies can store and/or retrieve all the information they need for all their trading relations? This would simplify and reduce the communications of detailed elements between companies. An early attempt to achieve this was successful in Germany where the SINFOS *data pool* was established. Indeed its success led to the German EAN organisation trying to persuade other countries to join their work on this to establish a single global data pool. What has actually emerged is a large number of data pools linked together within a global data network.

At the heart of this is a GS1 Global Registry. This takes a limited amount of information which can be accessed to open up more details about products. That information consists of the unique item numbers (GTINs), which have been at the heart of this book and global location numbers (GLNs) which were referred to in the last chapter. But the third element recorded is a set of global product classifications (GPC). In GPC products are divided into segments, which divide into families, which divide into classes, which divide into bricks and with many products in any single brick.

Does that sound familiar? It is nearly 16 chapters since I introduced the subject of BAN-L and the German desire 50 years ago to classify products. That led to a numbering system which gave significance to those classifications within its number structure. Whilst the current number system is non-significant, the Global Registry recognises that there is still a benefit in being able to focus on sub-sections of the worldwide inventory.

A second acronym that needs to be mentioned is VANs. If motor vans have been used for over a century to distribute physical goods and associated paperwork, then a new generation of VANs are now being used to distribute electronic messages. The concept is a simple one. If a manufacturer needs to communicate with a number of customers or a retailer with a number of suppliers, they can avoid the need to process all of these items of communication in separate 'envelopes' by putting all their 'letters' in one 'envelope' and transmitting it to a Value Added Network (VAN) for splitting and forwarding.

Finally, the development of personal computers, mobile phone technology and the internet has dramatically changed the way in which people communicate with each other. It has also opened up new possibilities in retail industry communication between companies, as well as with customers (which will be considered in the next chapter). Between companies, the EDI standards we have considered allow an alternative way of sending information in a different format to traditional paper orders, invoices, price lists etc. But there is now a further set of standards referred to as GS1-XML which allow companies to use the XML format to communicate actual documents.

This chapter was only ever intended to introduce the hidden world of electronic communication as it has impacted retailers. There is much more that has been developed. However, it is sufficient to note that a retail revolution has taken place in the last 30 years in this area. The specific way in which companies will have used these enabling technologies, standards and services will vary. Thus, one company may choose to send each message directly to another company, whereas another retailer may send all its messages via the clearing house of a VAN. One company may use large mainframe computers for processing and another will use the humble, but powerful PC. And the world wide web has allowed much smaller companies, who do not have computer staff and EDI expertise, to communicate too in this way.

Whatever the specific way of doing business, the overall impact on the products and services now made available to consumers as a result of them has been immense.

The same can be said of our next subject. For I now turn to a part of retail life which extends from the market stall I have referred to in Kenya to the largest hypermarket. It is as old as retailing itself but has been revolutionised by the introduction of item numbers and bar codes – for there are *Bargains, Bargains Everywhere.*

20. Bargains, Bargains Everywhere

I got into trouble in China. I had travelled there to chair my last EAN AGM in 2000. Following the meeting I was in a small town with my wife and daughter, about to cycle to a small village with our guide. She insisted that we must buy hats to protect ourselves. But she was horrified when I was prepared to pay the price asked by the stallholder. Her few words of English and many gestures made it clear that I must bargain to arrive at a final price.

This goes back to the experience I recorded at the beginning of this book in Ewaso Ngiro. It is clear that the process of negotiation and the desire to 'find a bargain' goes back a long way. It remains very widespread in many countries. However the form it takes in large scale retailing has changed. The element of negotiation has disappeared almost entirely but the emphasis on bargains is more pronounced than ever.

This struck me most forcefully when I first encountered the mass use of 'flyers' on visiting the United States in the early 1980s. The weekend newspapers were the thickest and heaviest I had ever seen. And much of the volume consisted of the colourful flyers of local retailers promoting their special offers trying to persuade you to visit their store rather than others this weekend for their short-term price reductions. Of course, this method of attracting customers was not new. But the overwhelming message was of *bargains, bargains everywhere*.

Of course, bargains are far from the whole story. Modern retailers have widely different strategies to attract and satisfy customers, ranging from those offering style or uniqueness at whatever price to those offering width of choice to those focussing on permanently low prices or specific bargains at the lowest price.   If scanning systems have allowed retailers to use detailed sales information to impact their decisions, they have done so within their specific broader marketing stance.

My experience in the UK in supermarket chains is that the use of bargains and special offers appears to have become much more widespread in recent years. Walking round a supermarket every area contains price reductions, buy one get one frees (BOGOF), buy two of a product or range of products at a lower price, buy one get second at half price, meal deals allowing a choice of products from three different categories at a fixed price etc. The significance of these activities is reinforced by recent figures showing that for some stores more than half the products passing across the scanners are involved in one form of promotion or another. Since such 'bargains' have become so influential in customers' purchasing decisions, I have chosen this part of the retail offer to illustrate the changes that have resulted to sales and marketing management and decisions from the introduction of article numbers and bar codes.

*****

In the early 1980s I was a member of a committee which reviewed the four-weekly Boots Special Offer programme. The selection of products for promotion was seen as important by all in the company as the spearhead for its marketing activities.

In its review, the committee looked at figures for the 'normal' and 'offer' buying price, selling price and sales quantities of selected products, expecting to see both much higher sales and higher profits from the reduced prices. After each promotion, the actual results were reviewed and lessons learned for future programmes. Two limitations were always present. All the figures could only be estimates because sales figures were deduced from stock counts before and after the promotion, and supplies to the stores during the promotion. And secondly, the figures only related to the products being promoted.

In this world, it would not be true to say that scanning provided the first opportunity to use actual sales information. Most major retailers had found ways of deducing sales from supplies and stock data prior to their introduction.

Some had gone further in trying to capture genuine sales data. For example, some had given short numbers to a limited range of products and key-entered these numbers at more sophisticated cash registers. Department stores often used a number system with Optical Character Reader pens at payment points. Shoe chains included a punched card in each box of shoes which was removed when a sale was made. Nevertheless this data was largely used for stock control purposes and was never available for the big food chains.

When Boots introduced scanning in 1986 a decision was taken to hold centrally the detailed sales of every product sold in every store (stock keeping units or SKUs). At the time, that involved discussions on whether it was practicable and cost-effective to do so because the storage capacity and processing power of computers was limited. Surely, it would cost less to summarise the data in some way? The problem there is that it would be necessary to know in advance what data to summarise and what data would be lost as a result. Since the number of stores installing scanning systems was small initially and the costs of computer processing and storage would drop substantially before widespread installation, it was agreed to hold data at a very low level.

Suddenly the problem was not an absence of real sales figures. Instead it was how to deal with the mass of figures and convert them into meaningful and actionable information. In trying to understand how this could be achieved, I will continue to draw from my own experience when bar code scanning was introduced in Boots in 1986.

Take the example of a money-off promotion for one size of one brand of toothpaste. The two problems in measuring the success of such a promotion had disappeared overnight. Firstly, where previous sales figures were estimates available after the end of the promotional period, now all the figures were accurate daily sales recorded as each item was scanned at the checkout. Not only did this give sales made of the particular product during the promotional period, but similar figures were available for the periods both before and after the promotion.

Management was aware for the first time of precisely what effect the promotion had on sales and profits of each product and on what happened to the sales level after the end of the special offer period.

And it was better than that. For secondly, the same figures were available for every other size and brand of toothpaste which was not being promoted. Thus the impact on sales and profits on other sizes of the same brand and similar sizes of different brands could be seen. So, a wider view could be taken on the effect of the promotion, and on the overall sales and profitability of either the whole of the brand or the whole range of toothpastes.

I use this as a simple illustration. Marketing decisions are made on the basis of many other factors including attracting customers into a store, competing with other retailers, introducing new products etc. But scanning coupled with supporting systems does provide a detailed factual basis for observing the effect of promotional pricing decisions.

Today, retail managers are not exploring new sources of information but are working with advanced systems which have been developed and refined over many years. They are able to develop sophisticated promotional programmes which are spearheaded by bargains but extend to other parts of their inventory. As a British Member of Parliament said recently, 'Supermarkets can base their pricing and promotion policies on sophisticated computer analysis of how we shop'. I will return to the rest of what he said in the final chapter when I ask whether this has proved to be a good thing.

Here are a few examples of the ways such analysis impacts decisions and policies :

- Introducing a new product - not only can the sales of that product be seen immediately but also the effect on the sales of other similar products can be measured. Thus new products can be subject to introduction to a limited number of stores at limited cost and extended rapidly if successful or withdrawn if unsuccessful.

- Managing ranges of products – the mix of different brands and different sizes can be managed to determine the optimum choice to be provided.
- Managing price – the effect on sales of changing the price of a product can be seen for both that product, other sizes of the product and other competing products. Retail chains can operate different prices in different stores as a policy if they wish, knowing that they have both the technology to manage such variability and the information to make changes quickly.
- Managing space – the effect on sales of giving additional space, a different shelf or a different location in the store can be seen.
- Managing seasonal merchandise – sell out rates can be observed and products reduced in price at an early stage if it appears that stock will be left at the end of a season.
- Managing store variations – the differing catchment areas mean that the experience gained from detailed sales data in particular stores can impact the choice of products in similar stores whilst allowing differing store profiles to emerge with different ranges.

This brief selection only lists what all retailers must do in their stores, whatever systems they use. But the immediacy of sales information puts them, at least theoretically, in the same position as my father was in his awareness of the changing rate of sale of all his products.

How the information is used in practice will depend upon the policies of a retail chain in terms of central and local control, the computer systems support in processing the data and producing decision-making information, as well as overall marketing policies. If all of that is not enough to take into consideration, another level of information is necessary to take the most effective decisions – profitability!

For I have written exclusively about the impact of sales data. Yet any retailer is ultimately interested not in maximising sales but in maximising their net profit.

This fact led Boots to develop a computerised profitability system in parallel with the sales information system. The Direct Product Profitability system established all of the costs associated with an individual product. The system recognised that the distribution costs, space requirements, customer service needs etc varied and that these costs need to be associated with sales information before product range and pricing decisions could be made.

*****

All that I have detailed has referred to retailers. Yet all of the subjects above are of equal interest to product manufacturers. They are concerned with the number of stores they supply, the space they are given, the promotional opportunities available and, of course, the price they can negotiate. So what difference has scanning made to them?

As we have seen, from the outset both retailers and manufacturers had been involved with the introduction of numbering and codes. Both were committed to their promotion and development so that their widespread use would bring financial benefits to all. But if retailers were to gain from the use of all this detailed sales data, would manufacturers also be able to share in the spoils? Here I must again go back to the history of EAN. This subject was the cause of much discussion and animosity in the 1970s and 1980s.

For retailers, the substantial costs of installing point of sale scanning systems was justified by not only the savings made in staff costs at checkouts but also by the other improvements to their operations from the use of article numbers as the basis of stock and sales management. For manufacturers, it was less straightforward. Their costs were associated with the relabelling of every product with an item number and bar code. They also anticipated cost savings from the use of a common item number in the assembly, despatch and invoicing of goods. But from the outset, they also saw the potential advantages of using actual sales data. In this area, their historic position was much worse that retailers.

For whereas retailers could deduce the immediate sales of items from stock counts and orders received, the only data seen by manufacturers was in terms of the orders placed by retailers. Of course, over a period of time the accumulation of these orders approximated to the volume of sales. But even then these figures were subject to any change to the number of stores selling a product and the space it was given. And it still only indicated their own sales but not the total sales of similar products produced by others. So manufacturers would supplement this by market research of customer purchases to determine their sales and their market share relative to competitors.

Recognising that all parties in the industry were dependent on each other to ensure the new system became established, early discussions sought to ensure that all would also benefit from its introduction. Therefore the initial EAN Memorandum of Agreement stated that basic sales data would be made available to the suppliers of bar coded goods at cost.

There were some matters undefined within this agreement including what was meant by 'basic data', what was meant by 'at cost' and what mechanism would be established to provide the figures.

The desire of companies to obtain this new information can be seen in the minutes of meetings during the early 1980s, well before any significant number of scanning stores were in place. The annual EAN General Assembly of 1981 gave some time to the subject. The UK organisation reported that they had appointed Neilson, a leading market research company, as a clearing house for data and that they had defined 'basic data' as any supplier's own information plus the total figures for any product groups represented by those products. In other words, the toothpaste supplier would get a high level view of the sales of their own brands and the total sales of toothpastes. However, individual retailer figures and those of competitors would be protected and hidden within these industry-wide statistics.

I have mentioned the UK initiative because, as a member of the Article Number Association (ANA) Council, the EAN body in the UK, I was able to see the slow progress of this initiative throughout the 1980s. In practice, retailers were very cautious about releasing their own sales data. They were all concerned that information was not divulged which disadvantaged them in their competitive position with each other and in their negotiating position with their suppliers. It was soon clear that an open world where a substantial amount of data was going to be released was not going to happen.

So has the world changed at all for their suppliers? In some ways, they have much better information than they had historically. The improvements in stock management and logistics systems have meant that the relationship between orders and sales is much closer. And retailers are anxious to achieve high service levels, giving them an incentive to make their suppliers aware of any fluctuations in customer demand for their products. But beyond that, the provision of more detailed sales information is simply one element of the overall trading relationship and negotiations between individuals. That varies between companies, market segments and countries. Overall it could be said that manufacturers have gained from the availability and timeliness of sales information, though perhaps not in the way they envisaged when drawing up the Agreement nearly 40 years ago.

*****

To conclude this chapter I look very briefly at a subject of interest to both retailers and manufacturers – who are our customers and what do they buy? Surely the sales and profitability of every product in every store is enough data to process. And yet when I left Boots in 1992, I assumed that this was simply the beginning and that another revolution was about to take place because of scanning. Immediate information had become available about the sales of every product. But if individual customers could be identified with each purchase, then the next level of detail would be immediate information on the sales of every product to every customer.

211

It took longer than I expected but the presence of store cards and the very low costs of computer storage and processing means that this level of detail can be used to target marketing directly at you, the store card holder.

And you will have seen that it does happen. Perhaps you have been given a money-off voucher for your next shop. Since you seem to get those more often after a period when you have not shopped with a retailer, you can assume your past shopping history has been recorded and this is an attempt to get you back into a store.

Specific purchases and actions taken by you have been logged and are being used. On the internet, you will know that any purchase you make is likely to be followed by a stream of marketing communication. My daughter's experience is typical. The Amazon marketing system assumed that her fun purchase of a bee finger puppet for her husband was significant. So she spent six months deleting their emails for other finger puppets. But you do not even need to make a purchase. My innocent enquiry for a friend on a train website to find the price of trains to High Wycombe led
to advertising appearing later on unconnected websites which had clearly made use of my particular enquiry.

For manufacturers it means that data is available on what particular groups of customers are buying their products and what effect changes of price and promotional activity has on those groups. Of course, the provision of that information is again part of the trading and bargaining relationship between trading partners.

To conclude this chapter, I simply observe that my title 'The Iceberg' for Part III of this book is entirely appropriate. For what you see at the checkout is only a small part of the scanning revolution. That is only the tip of the iceberg. Below the surface the data from the reading of that bar code is used by retailers for their promotional programmes and their bargaining relationships with their own suppliers. Furthermore, when allied to your personal data it has changed the ways that you are targeted.

21. They've Changed It Round Again

I nearly didn't write this chapter. It was present from the first sketch I made of the subjects to be included but it almost got removed at the last moment. My elder daughter changed all that. It was a simple and innocent request to take her for a quick supermarket shop. (Is a quick supermarket shop the perfect modern example of an oxymoron?) We would only be gone for an hour, she said, because she had two fairly short shopping lists for herself and a housebound lady she helps. We came back two hours later. For as we entered the store, she exclaimed, '*Oh no, they've changed it round again*'. Her sense of frustration led me to reinstate this subject in an abbreviated form, for if the book is trying to indicate the impact of item numbering on the shopping experience of customers, it needs to be included.

Don't get me wrong. Changing the organisation of products in shops has always occurred. After all, a retailer has strictly limited space and must always seek to find the best ways of using that space profitably. Some of those changes relate to seasonality – Christmas, Easter, Halloween, Mother's Day, Father's Day etc – where additional space must be found and other areas of merchandise may be adjusted as a result. Many of the changes relate to sales promotions, where specific areas are constantly featuring different products. Other changes are more major and reflect the decision by a retailer to introduce or remove complete product areas, and to respond to changing competition. So in some senses change is simply business as usual.

Yet I am somewhat more cynical than I used to be on the subject of change. That is somewhat strange when I have spent most of my working life advocating and implementing radical change and persuading reluctant colleagues of the need for such change. So perhaps it is simply a question of the ageing process. But no, there is more to it than that.

The pace of technological innovation, the shortening of the life cycle of products, the need to gain competitive advantage before innovation is copied and the availability of information on which to make changes have combined to make the ongoing accelerating pace of change excitingly inevitable. On the other hand, several other common factors have combined to raise doubts about the need for the degree of change which actually is taking place. The move from long-term thinking and action to short-term decisions and judgements, the frequency of change in the personnel who manage processes, and the pace of life which either cannot or does not give time for one change to be effective before it is judged and superseded all lead to even more change.

The combination of all of these factors has led to spectacular developments which have transformed our lives. But as the financial crisis has demonstrated, new individuals developing new ways of doing things, unsympathetic to past ways of working and long-term considerations can also have disastrous effects on our lives too.

So, to come back down to earth, why do retailers change their shops around so much when they know that it will provoke an adverse reaction from many of their customers?

*****

I don't know how James Ritty managed the space in his store. But I did see my father's greengrocery shop over a period of years and it experienced very little change in the products he sold or in the way they were displayed. The seasonal changes he made in fruit and vegetables were probably greater than those seen today because there was no availability of non-seasonal products, which today can be flown in from anywhere in the world. But there were relatively few new products because they were simply not available, as the change between cookery books and cookery programmes on television will indicate. In larger shops and retail stores, much greater change took place and Boots was typical in producing weekly product announcements and allocations of new products.

The buyers for a retail chain today have even more suppliers and products seeking to find space in their stores and will be taking decisions on adding and deleting products from the inventory every week. The process of taking those decisions and of introducing new stock will vary. But all will be faced with finding space for each new introduction.

The development of computers allowed such retailers to produce *planograms*. These were plans which might simply show which groups of shelves should be allocated to different product groups but could also show the precise location and space given to individual products. Today you can observe in a store the effect of such planograms by looking at shelf edge labels. Earlier we noted that these were required to give the description and price of a product because price stickers had been removed from the products themselves. But it is likely that you will see additional information given on the shelf edge to indicate the number of outer packs of the product to be stocked and possibly even a coded location in the store. This information can be used within the store to replenish the shelf, either from a visual check of the stock or by scanning the shelf edge label and recording the remaining stock on the shelf.

But the space information can also be made available to the buyer of a product range. The space allocated in each store is known, which allied to sales information can assist with decisions to change or eliminate the space required for any product. Furthermore, as we have seen in earlier chapters, additional information is readily available on the size of the packages of products to make changes much easier to manage. Given the detailed information potentially available on sales trends and space, and the relative ease with which new planograms and shelf edge labels can now be produced, it is not surprising that specific product changes are frequent. Therefore at the lowest level I must accept the sudden disappearance of the brand of coffee or shower gel I came to buy.

I will also find that, even if the product is still stocked, its location within the product area may change.

The power of computers means that it is possible to analyse in ways not previously possible the impact of placing products on particular shelves or in proximity to other products.

And that level of detailed information leads us to decisions on more major changes made within a store. There is now a vast amount of experience which has been gained in the organisation of supermarkets and the relative positions assigned to different product categories. This could mean that major changes should be less frequent. However, it is also relatively easy to experiment with different formats in an individual store and analyse quickly and comprehensively the effect of the changes made.

So what should you think when you cannot find the products you are looking for and feel like shouting, *They've changed it round again'*. Perhaps your understanding of all the information now available to retail managers will lead you to conclude that they are using that knowledge to make your route through the store easier and in 'improving your shopping experience'. In this way they will hope to retain you as a customer on a regular basis. Or you may feel that they have found new ways to persuade to increase your purchases and their profits. If you are somewhat more cynical, you may wonder whether that is simply a case of new personnel who must do things differently to their predecessors.

As I mentioned at the start of this section, one of the earliest retail lessons I learnt from my father was the importance of retaining customers and the ease with which they can be lost. As a result, I would love to think that all of this investment of time and money in the supporting systems I have described in this book has led to a greater focus on customer service. But you will know from my introduction to this chapter that I would not be surprised if there are times when you wonder whether all the changes were necessary or desirable.

*****

Before moving away from the world of retailing, I should extend the subject to the area of the greatest change and most dynamic growth in the past decade, that of internet retailing.

Part of the retail market has for a long time involved mail order purchases. But the development of the internet has allowed non-store retailing to flourish. Its impact can be seen in all types of retailing, not simply in the traditional areas occupied by mail order specialists but extending in such a way that most retail store chains have felt that they must have an associated internet offering.

As I write this, two parcels have just been delivered for my daughter which are the result of her internet ordering. I note that both have a bar code on the outside of them and that, inside one of them, the order number is given in the form of a bar code. I also realise that the second parcel has been sent as a replacement because an earlier one had been sent here with the correct address on the label but the wrong product (for a different customer) in the package. So, whilst this internet world has also been impacted by the introduction of bar codes, it does not mean that it has resulted in an automated activity where individual human mistakes cannot be made.

The order you will place on the internet will be in terms of a product's description, probably accompanied by a photograph of the product. But every aspect of the processing of that order will probably have benefited from the introduction of item numbers and bar codes.

If your order is from a supermarket chain for home delivery, it will have used the GTINs associated with the products you have selected. If you have used Amazon for the delivery of a book, a conversion will have been made into the ISBN/GTIN.

Whether you are dealing with the internet site of a retailer or a specialist supplier with no retail presence, it is probable that your order, their warehouse operation and whatever delivery system they have used will use both numbering standards and some form of bar code.
Indeed it is not unreasonable to say that, without the development of all of these standards and the ability to convert products and locations into numbers, this form of retailing could not have developed in the way that it has.

PART IV – Beyond the Retail World

22. How Do They Manage That ?

My love of numbers and of mathematics helped me to understand one important issue in family life. There is a fundamental difference between having two children and five children. With two children you must cope with the fact that they may argue with each other. However, with five children there are not just five but ten different pairs where an argument may start. Fortunately, though sadly too, the days of resolving such matters as who will sit next to the windows in the car have long gone.

So how does GS1 with over 100 different member organisations survive and prosper? How do they manage a global system and standards? It does not need such a high number to produce fundamental disagreement and animosity, as global politics shows us every day. And in this world of numbering and codes it needed only two. For this global system might not have happened and the global organisation GS1 might never have existed.

*****

On 10th October 1980, Albert Heijn as Chairman of EAN wrote to Robert B Wegman as Chairman of UPCC as follows :

'I very much fear that ..... we will have to face the fact that the UPC and EAN systems will no longer be mutually compatible and therefore must agree to operate independently of one another.'

In Part I of this book we saw how article numbering, bar coding and scanning developed in the United States. We also saw how a group of Europeans took these ideas and modified them to produce a compatible system for use outside North America. Part II has shown how this basic system works and how it moved outside North America and Europe to include most countries in the world. Part III has indicated how it affects almost every aspect of the global retail industry, and in ways that were never foreseen originally. And Part IV will show that its use has gone into other sectors than retailing.

A vast amount of commerce in terms of the movement of goods and information is now dependent on these systems. Any problems with either the underlying global standards or the ongoing global management of the systems would have major practical and financial consequences. So the benefits of having a single global system seem so obvious that its realisation was surely inevitable. But that was not the case. Achieving such international consensus and management has been far from easy, as the quotation above indicates.

*****

When we left the embryo structure at the end of part I, the North American system was managed by the Uniform Product Code Council (UPCC) and encompassed the United States and Canada. The European system was managed by European Article Numbering (EAN) with organisations in each member country.

As the latter system expanded to other countries and continents its name changed to become EAN International, but it retained the same structure of country organisations managing the system in their own country and meeting together as a General Assembly to control the system. All were set up as not-for-profit organisations and managed by a combination of senior representatives from retailers and manufacturers.

Whilst EAN continued to expand its geographical spread, the independent UCC system continued to develop its systems and standards in North America. The organisations were in contact with each other but the tension between them is reflected in Albert Heijn's comment that the two systems must go their separate ways and that the future world would not be that of a single global system seen today.

So what was his particular reason for believing this in 1980? I return to the Twinings product mentioned earlier which had a 12-digit UPC number rather than a 13-digit EAN number. I noted that this was caused by the fact that the the earliest scanning systems in the US were designed to read 12-digit numbers.

221

Albert Heijn and those leading with developments in Europe recognised from the outset that systems and scanners in Europe would need to be designed to handle not only their new 13-digit EAN numbers but also the shorter UPC numbers, because stores would be selling products imported from the US with a UPC number.

But they also foresaw a major problem and Heijn wrote repeatedly from 1976 onwards to the US about it. Unless US systems were modified at this very early stage of development to read both UPC and EAN, any exporter to the US would need to number and label products differently for the US market than for all other countries. The EAN system had been designed so that any product could cross country boundaries with its unique number and did not need to be re-numbered outside its country of origin. But this fundamental would be completely undermined unless US equipment and systems were modified. Surely, in the interests of a global system, this could and should be achieved?

The summary by Tom Wilson of the situation in June 1980 to Andrew Osborne of the UK organisation indicated why no progress had been made in the previous four years.

'None of the original US scanner equipment was designed with the EAN symbol in mind. Thus, when that variation of the UPC code was adopted in 1977, equipment manufacturers in the US had to initiate changes in the decode logic and software of their in-store scanner systems, in order to allow US retailers to purchase systems with EAN reading capability. In addition, because US retailers had set up item files with 10-12 digits, they were faced with restructuring them to accommodate the 13-digit EAN format. Because there have been very few European scanner installations, European manufacturers with significant export volume to the US have naturally tended to mark their products with UPC rather than EAN symbols. Thus there has been little if any incentive for US retailers to seek EAN reading equipment or to restructure their item record files.'

However, it was the final paragraph of the letter, copied to Heijn and Wegman, which incensed EAN and led to Heijn's letter to Wegman :

'Given the voluntary nature of the UPC standard program and the US antitrust framework, the capabilities of the UPC Council to engender action on this issue is quite limited. However, we continue to believe that a gradual move toward EAN reading capability is taking place in the US, and will continue to do so. At the same time, it is likely to be a number of years before food exports to the US can be marked with EAN symbols and expect universal acceptance.'

However, Stephen Brown's book on the history of UCC has summarised what this meant in practice :

'The UCC Board of Governors never really understood the EAN position, viewing it as a tempest in a teapot. Thus, although the Board adopted formal positions that equipment and software should be capable of scanning both UPC and EAN symbols, very little energy was put behind the effort.'

Unfortunately the words of Heijn were ignored and so, even today, over 35 years after he first raised the issue, dual numbering is still necessary for many products. The unnecessary extra costs have been huge, but despite Heijn's threats, businesses have been forced to accept them in order to sell their products. However it is not surprising that some companies such as Twinings have chosen to use a single 12-digit number, which is accepted worldwide, rather than a 13-digit number, which may be rejected by a US retailer.

This correspondence is only quoted here because it indicates how easy it would have been for a single issue to lead to a parting of the ways and potentially independent and competitive systems to involve. That did not happen even though no meaningful action was taken by UCC to address the issue until the early 2000s.

But the issue did affect relations between the two organisations for many years, even though they worked on many technical issues together. Somewhat closer links were established in the 1990s through the efforts of the permanent officials of EAN (Etienne Bonnet and Reinhold van Lennep) and UCC (Hal Jackett and Tom Rittenhouse). The chairmen of the organisations also attended each others meetings and met together. Whilst this was happening at a personal and organisational level, at a business level the rise of multinational organisations was leading their business representatives to seek even closer relations between the two bodies. The determination to change existing structures in the new millennium was pushed forward by Tim Smucker as Chair of UCC and Laurie Wilson from Australia as President of EAN, with the particular practical help of Brian Smith from Australia. This resulted in the coming together of the two organisations to form a new global organisation GS1 in 2006, with Board members from all over the world. Within the new structure, the EAN country organisations continue to exist and in the US GS1US has become a similar country organisation successor to UCC.

Ironically an early President of GS1 from the US, responsible for resolving outstanding issues and ensuring that such problems do not occur in the future, was Danny Wegman, the son of Robert Wegman, to whom Albert Heijn's angry letter had been written a quarter of a century earlier.

*****

If the above history indicates how two independent organisations were avoided, there were also early threats to the very nature of EAN and the vision of the founders. We saw earlier that the EEC took more than a passing interest in the embryo organisation. In October 1976, it was reported :

'Mr Besnard (EEC) had clearly stated his position : he envisaged EAN as a lightweight organisation and as a de-facto organisation ....of a 'para-governmental' nature, that is to say one in which the State has a majority voice in the sense that central decision-making is directly influenced by the State in each of the Common Market countries belonging to EAN.'

This was completely at odds with the concept of a business-led organisation free from such political interference. And EAN resisted the EEC position, eventually persuading it that the requirement for open trading across country boundaries would be met without the heavy, delaying, political hand of governments being involved. Who knows what the outcome would have been if the state control indicated by the EEC had been accepted. Is it too malicious to suggest that discussions on the controlling organisation with its voting rights and fee structure might still be going on today without a bar code in sight?

But even with the structure which was established and the overwhelming desire to reach consensus and agreement on a sound technical, business and financial basis as quickly as possible, tensions arose.

The minutes of EAN meetings hint at these tensions, and those attending confirm them. Matters came to a head at a meeting in London in May 1977. Albert Heijn insisted that the minutes of that meeting must be changed.

The following sentence has to be modified from 'On Health grounds, Mr A Heijn leaves the meeting and asks Mr J Vermaelen to take over the chairmanship' to read 'Mr A Heijn asks Mr J Vermaelen to take over the chairmanship and leaves the meeting'.

The reasons for this change were provided to me by Albert Heijn himself and confirmed by three others at that meeting. Heijn stormed out of the meeting at the end of the first day indicating that he had had enough of the organisation and was flying back to his business in the Netherlands immediately. The reasons for his anger are only obliquely reflected in the minutes preceding his departure

'A lengthy discussion takes place concerning the organisation of EAN and different remarks are expressed about the proposals made.'

For Albert Heijn, one of the leading retailers worldwide, used to making major decisions concerning his retail business, the protracted discussions and entrenched positions were more than frustrating. He had become involved because he believed major business benefits were possible for retailers and their customers by the new standards and technology. But the lengthy process of achieving that had become too much for him.

He was persuaded to return to chair the following meeting ..... and the meetings for the following ten years.

*****

These matters from the past serve to give some understanding of the world of standards. It is very different from the normal business world. It does not have a Chief Executive who can make unilateral decisions. It is far more concerned with the long term than in producing short term results. Whilst it does have voices which are louder and more persuasive than others, it must find ways to reach consensus decisions which are not unreasonably delayed but which will stand the test of time.

Given that, how in practise does the GS1 organisation operate? How are standards developed, agreed and implemented? Who manages, promotes, controls and provides education in the use of agreed standards? And, how is all of this activity funded?

As we saw in Part I, both UCC and EAN were formed by the direct involvement of leading figures in the retail industry from both retailers and manufacturers. In other words, from the outset, the driving force behind them was neither government nor technology-manufacturer based but industry based. That has remained the driving force and principle throughout the lives of UCC and EAN, and of the current GS1.

The structure of GS1 and of individual country organisations is for a Board of Management to be composed of retailers, manufacturers and other key industry representatives.

The boards will reflect the different way in which industry is organised in different countries and the importance of particular non-retail sectors in a country. These boards take decisions in their own countries concerning their priorities. However, where global decisions and standards are required, their voice is heard through their representatives to the world GS1 body. All of these country organisations employ permanent staff to operate them, and most of the detailed work both nationally and internationally is done by them. Much use is made of the technical and business expertise of companies throughout the world in producing new standards to meet new requirements. Such standards and symbologies will often originate outside the GS1 organisation and be taken or adapted for GS1 members.

The achievement of this historically posed more of a problem for EAN than for UCC. Firstly EAN had to handle the development in standards for existing members whilst introducing them to a growing number of new member countries. As an example of the tensions this caused, I can recall as chairman of ANA (the UK EAN organisation) bringing a proposal to add the arrow sign often seen at the end of the printed article number. Our research in the UK had clearly shown that many scanning misreads were caused by insufficient space left at the end of the printed bar code. To reduce these problems the UK wanted to change the way the number was printed in all countries. However, that international meeting in Copenhagen in the early 1990s included countries who claimed they had no such problem, and others who had only just introduced scanning and did not want to go back to their manufacturers to ask them to change their recent product printing of bar codes. So, although the meeting allowed the UK to introduce the change it wanted as an option, it did not feel it necessary or agree to it becoming a new global standard. This did not produce the best result for the UK because it meant that more reading errors might occur from imported goods. But it was a pragmatic decision which did not have unacceptable consequences.

In addition EAN had to deal with many languages. This involved simultaneous translation facilities for English, French and German at meetings. In 1988 the annual meeting was held in Japan.

After considering the options and costs of translation, together with representations from Spanish speaking countries that their more widely spoken language should also be used, it was decided to use English only. Individual countries used their own translators if required. But although English has continued to used for such meetings, all standards must be translated accurately and unambiguously into each member's language.

Much of the work of GS1 country organisations today is associated with the management, education and promotion of existing systems, for new individuals and companies come to the world of article numbering every day. Their finance comes from fees charged to each member company requiring a number. In all countries this takes the form of an annual fee for the use of their unique company number and support in its use. When a company ceases to exist, that number is returned to the country organisation for potential reuse to a new member. Before the formation of GS1, UCC had a different system where a single licence payment was made for the use of a number by a company. GS1 US now uses the annual payment method to finance its work. All country organisations are self-financing but non-profit making, paying fees themselves in turn to the global GS1 organisation to meet their costs.

In summary, it is a simple structure allowing as much autonomy as possible at a local level and using local knowledge and expertise, whilst taking corporate decisions and giving global support where that is necessary. Its effectiveness and reliability can be seen every day by you as the products you buy are uniquely identified. Simple but really quite remarkable and worthy of some element of praise in the achievement when saying, '*How do they manage that?*'

23. Bar Codes Are Good For Your Health

I appear to have picked an appropriate week in which to write this particular chapter. As I start it, I am confronted with a set of newspaper headlines :

'Horse in School Dinners' (The Sun)
'We've Been Eating Horse for Months' (Daily Star)
'Police called in to investigate 'criminal' horsemeat scandal' (The Guardian)
'Now Tesco admits its bolognese is made of horsemeat' (Daily Mail)

It is the latest food scare with products in at least 17 countries said to be involved. Frozen meat products have been withdrawn from shelves in the Netherlands, Belgium, France and the UK. There is a growing war of words between supermarkets, processors and producers over who is to blame. The Environment Secretary has sought to reassure the public and the Chief Medical Officer has felt it necessary to say that, *'there is a very low risk indeed that it would cause any harm to health'.*

But food safety is not the only issue to have dominated the headlines it the past week. An earlier theme was hospital safety :

'NHS : No one is safe' (The Times)
'NHS In Crisis – 1,200 died in their care' (Daily Mirror)
'3,000 more patients have died needlessly in hospital' (The Daily Telegraph)
'Minister calls for police enquiry into NHS' (The Daily Telegraph)

This follows a report on problems in one hospital but is once again part of a subject which goes beyond one incident, one establishment or one country. Sometimes single incidents receive a lot of publicity. A few years ago the actor Denis Quaid was involved when his new-born twins almost died after being given the wrong drug which was 1,000 times stronger than the correct one.

This led him to research the subject and suggest that, '*100,000 people are killed every year in hospitals by a medical mistake*'. The vast majority of incidents are either not recognised or not reported. The medication my wife received last week indicated it should be taken twice daily, though the doctor had told us it should be twice weekly. The pharmacist admitted making an error on his printed label but, since we did not report the matter, it will not be included in any statistics on mistakes made in pharmacies.

*****

Perhaps at this late stage of this book I can reveal an unexpected passion. In the retail industry the comprehensive, integrated use of bar codes has driven improvements in business efficiency as we have seen. In particular it has reduced the number of errors made by checkout operators, warehouse and office staff by changing the way they worked. Surely the same approach could be made to issues of health and food traceability (from farm to plate). Surely there must be an opportunity to improve safety standards but even more importantly to save lives! Of course human errors occur, but can we use this technology to re-engineer processes and either eliminate or significantly reduce the possibility of human error?

Back in 1970, as recorded in Chapter 4, McKinsey were asked to look at the economic case for fundamental change to the retail industry. It was one of the factors which influenced business leaders to make the commitment to finding a radically different way of working. In October 2012 the same company produced a report on the health industry, '*Strength in Unity ; the promise of global standards in healthcare*'. It is worth quoting from their summary :

'Imagine a world where a patient's records capture the brand, dosage, and lot number of each drug and medical device she uses, along with the name of the physician who ordered the product and the nurse who administered it ; where bedside scanning confirms that she gets the right product in the right dosage at the right time ; where hospitals and pharmacies know the exact location of short-supply medical devices and drugs and when they can be delivered ;

where regulators can recall adulterated products with accuracy and speed from every point in the supply chain ; and where manufacturers can monitor real-time demand changes and shift their production schedules accordingly.

In this world, patients would enjoy consistently safer and more effective healthcare, with fewer mistakes and shorter average hospital stays. Redundant activities and costs would be driven out of the system – reducing the cost of healthcare to society .... doctors and nurses could spend less time with paperwork and more with patients....This world is technologically possible today. But it has yet to become a reality because the healthcare supply chain, from manufacturer to patient, remains fragmented, with limited visibility and interconnection'.

That is much easier to write that to achieve. And so much is already being done in the health service. But perhaps, just perhaps, we may be on the brink of a health revolution. And perhaps a book like this will be written in years to come of how that was finally achieved.

Of course, the world of healthcare is very different and much more complex. I described earlier my experience of introducing scanning to Boots stores. The experience of the UK National Health Service in trying to develop integrated computer patient databases has been far more difficult and produced far more problems. The following lists some of these differences to introducing new global ways of working :

- Unlike retailing there is no single process like the sale at a checkout which is a driving force for change. However like retailing the inevitability of human error and need to find ways of reducing their incidence is fundamental.
- Unlike retailing there has been no direct equivalent to the single driving force of the joint group of key retailers and manufacturers. However there are major pharmaceutical manufacturers and there are some strong buying groups.

- Unlike retailing much more than a simple number structure and symbol is required to address the range of health service activities. However this has been recognised and the development of reduced space and two-dimensional bar codes has provided a basis for change.
- Unlike retailing no single legislative change (such as new food labelling requirements in the US) has coincided with a fundamental system change which allowed implementation at a marginal cost. However international legislative forces will be significant as patient safety is given focus.

During my period in the 1990s on the Board of EAN, the healthcare industry was the subject of constant discussion. I assumed when I left in 2000 that it would be using systems similar to those of retailing by now. There has been much development and there are initiatives around the world but progress has been fragmented and uniform global systems remain a distant goal. My optimism may now be somewhat less but my belief and passion remain that a global solution is possible, if difficult, and has the potential to transform the outlook for patients and reduce potential costs for the future.

*****

A similar description and conclusion could be written about the subject of food safety and traceability. The food scare I referred to earlier is by no means the most serious and is unlikely to affect people's health. But in the UK at least it has triggered a wider debate about the management of the complex world of food processing, whether that food is bought in shops or restaurants. It has highlighted the complex nature of seemingly ordinary products such as burgers and lasagne in terms of their ingredients and the sources from which they come. Maps in national newspapers have shown the number of countries which might be involved in the purchase and procurement of a single ingredient and alerted the public to the fact that the source might change from week to week.

This is a world every bit as complex as that of the health service. It is also a world which affects the safety of all of us. Indeed the modification of food products with preservatives and additives has already affected our health in ways which will not really be understood for many years to come. The latest health scare does indicate that tracing what has happened to all the products can be done – eventually. But that is far away from the vision that McKinsey might have provided if their report had been about food rather than health. You can imagine that they might have suggested an integrated world in which almost at the press of a button any single issue with a processed food product leads back to its source and then forward to other products with material from that same source in a very short time.

To achieve this we must recognise a key requirement which the retail industry did not have to address. A particular sized pack of Kellogg's Corn Flakes only required a 12 or 13-digit GTIN. This number was sufficient to point to all other information about the product which was held on computer files. For full traceability, it is necessary to identify some other details such as a production batch number or date about an individual item. However, the technology, systems and standards are now available to do this and are partially in use. The ability to record a lot of detailed information in a small space with newer and smaller two-dimensional and databar codes has made this possible. However the costs and difficulties of moving to the use of such standards and installing scanning equipment able to read such codes cannot be underestimated. Whilst the consumer might want full traceability, there is an even greater demand for inexpensive food. Legislators and the industry will be mindful that these are not always mutually compatible.

*****

These are only two industry sectors where standards based on the GTIN and bar code are currently in use and will be expanded. There are many others.

For example, the logic which has led to the use of numbers, symbols and associated data held on computers in the retail logistics process applies equally to any goods which are transported. Of course some large global businesses have developed their own systems. But the problem for most businesses is much the same as that experienced in Germany and France in the 1960s. If many different businesses want to send goods to many other different businesses, they are forced to communicate in different ways using different systems unless a common global system has been agreed. These countries were able to be involved in a solution which allowed them to retain their own numbering systems within the new European EAN system. Not all industries and all businesses are as fortunate as that, making the process of change to what is recognised as a more holistic system much more difficult and costly to achieve. Nevertheless the standards referred to in this book have been adopted in many countries in many business sectors.

One area which is in its infancy is what is referred to as B2C (business to consumer). This can be seen if you use a mobile phone to scan a bar code and are able to find some details of the product. Such a service can be provided by making use of the databases which hold information or images about a product. The current use of such services is small and the anecdotal reports of the wrong product information being retrieved are many. Your newly gained extensive knowledge of item numbers will appreciate that such systems do depend on every product having its own unique number. You will also know that, although this is normally the case, there are a range of numbers referred to as 'restricted circulation numbers' which are designed only to be used in one retailer's stores. Thus different supermarket chains may have used the same number and the wrong product details may appear if one of these numbers is scanned. To fully implement B2C means that these numbers are no longer restricted and such retailers would need to consider renumbering with unique numbers.

This is one example of the way in which changes in technology are forcing a reconsideration of the system and standards which have been in use for the past forty years. So it is appropriate to move on to a much wider consideration of that subject and what the future may hold for the world we have described. Are we about to see a period of gradual evolution or be subject to another retail revolution?

24. And The Next Number Is

'To meet the demands of the fast-changing competitive scene, we must simply learn to love change as much as we have hated it in the past. Our organisations are designed, down to the tiniest nuts and bolts and forms and procedures, for a world where tomorrow is today'

In 1987, a US management consultant named Tom Peters wrote those words in a book entitled, 'Thriving on Chaos – Handbook for a Management Revolution'. It was published shortly after the first few Boots stores had installed scanning systems. Within the company, EPOS was a revolution. Operations within the stores were being transformed, buyers were beginning to use information they had never previously seen to make decisions, and my staff were trying to manage the acceleration of the installation programme. Part of the title of the book, 'A Management Revolution' seemed particularly apt, though I hoped 'Thriving on Chaos' was not. Yet this massive change was being inflicted on a retail chain that, despite growing competitive pressures, was not failing but had an outstanding record of profitability, highly regarded by its customers. So the opening words of Peters' book were encouraging:

'There are no excellent companies. The old saw, 'If it ain't broke, don't fix it' needs revision. I propose, 'If it ain't broke, you just haven't looked hard enough'. Fix it anyway.

In the past 20 years, the pace of change has quickened and continues to accelerate. Whether I read and hear about commercial companies or public services, the message is the same. It is of constant change, with one new initiative barely introduced before another follows on. To those subjected to these changes, the cry is often, 'is it really necessary, enough is enough, let us digest the previous change and make it work'. To those proposing the changes, the louder cry comes back, 'it is necessary, enough is not enough, we must change'.

Now, I must be getting older because, whilst I can still read the words of Peters and shout '*Yes*', I also find myself muttering '*Yes ....but*'. For, on the one hand, the world of today has a global dimension, is more competitive and products have shorter life cycles It requires innovative solutions, introduced more quickly and managed more effectively than was the case twenty five years ago. But, on the other hand, the world is also one of decision makers who are often more remote from their customers, have spent less time in their organisations and who have the siren voices of management consultants and city analysts (with different agendas) urging change on them. We cannot conclude that the highly-rewarded advocates of many of these changes have always produced better results in the public or private sectors for their stakeholders. Indeed the experience in the finance industry of overturning a decades-old business model focussed on traditional bank lending with more innovative products, for example, might lead to the argument that many changes were not thought through properly and some were even unnecessary.

In that spirit, I approach the future of the subject of this book. Where does this system sit in this world of change? With a life span of nearly forty years already, should we be looking for something new?

So, will I look at this world radically advocating the overthrow of numbers and codes and their replacement with a revolutionary new system? Or will I more cautiously argue against any change? And, who am I, with so many hours invested in the current system, to provide any objective or thoughtful view. I leave you to judge as I seek to look ahead to say, 'and the next number is .....'

*****

Our starting point must be the <u>number</u>. Has it got a limited life expectancy? To answer that I go back to the reasons for its existence. In France and Germany, the retail industry wanted a way of uniquely identifying products and found it in terms of a numerical structure. Later, in both the United States and Europe, a similar desire led to globally unique numbers.

These numbers were sufficient to provide a key to access any other information needed about that product, whether at a checkout in terms of price and description, or in offices and warehouses where computers held other details. During the last forty years questions have been raised on whether there would be enough numbers to satisfy the growing demand for them and whether the numbers were of sufficient length. But there has been nothing that has led to any overwhelming requirement for their replacement.

So the Global Trade Item Number (GTIN) continues to do what it says on the label. It identifies items in a globally unique way and has become the global language of business. Learning a new language is not something to be undertaken lightly. Any fundamental change would have profound implications for all computer systems using the number and would be more difficult to justify as a result. But that is not the basis for my view that it still has a long shelf life. I believe that if it did not exist and a global solution was being sought today, it is likely something similar would be created. It might not be the same length, have the same contents or be wholly numeric, but it would be similar. In the future there will be further new elements of information which need to be identified, particularly as the use of this number system continues to extend into other non-retail sectors. But these will be accommodated without the need to abandon the GTIN.

In the north of England my parents' generation had an expression when they bought a product in later life which they did not expect to replace. They said, 'This will see me out'. My first prediction, albeit as one who has celebrated his seventieth birthday, is that the GTIN will not be replaced in my lifetime, it will see me out.

*****

Can I make the same prediction about the bar code? Again I go back to the reasons for its existence. In Philadelphia 65 years ago the search started by Joe Woodland was for some symbol which would result in retail customers being served more quickly and at less cost. In the US and Europe the later search which resulted in today's bar code was designed to achieve the same end.

238

It was one way of representing the number such that reading the symbol would lead reliably to that number.

Again I ask the question whether it would be created now if it did not exist. Despite the simplicity of the bar code and the ease with which it can be printed on so many types of item and surfaces, I think my answer must be negative.

But to consider what could be created instead, I need to return to the problem it sought to address, checkout queues.

On the day before writing this chapter, I was in a supermarket. I waited for some minutes at a checkout behind three other shoppers. My goods remained in the shopping trolley for some minutes before I was even able to place them onto the checkout conveyor. The store had a bank of checkout self-scanners but there was an even longer queue for their use. So the issue of queues and delay has not gone away and remains frustrating for many shoppers. As for those responsible for the management of such chains, the long term search goes on to reduce queues and staff costs.

When Joe Woodland looked at this problem he started with a blank sheet of paper, or rather a bare stretch of sand. If we do the same with not only today's technology but an assumption of future developments too, what solution might we find? The starting point could be the complete elimination of checkout staff and even the checkout as we know it. Could we identify all the products purchased as they are selected and simply pay on exiting the store, perhaps with some form of security check? In fact this type of system is already in use in a limited number of stores. Shoppers use a bar code scanner attached to the shopping trolley to self-scan rather than wait to scan at the checkout. That would seem to indicate that no alternative to the bar code is necessary. But let us consider a couple of examples of alternative technologies.

The first is some form of electronic tag. Both the technology and the standards already exist. A Radio Frequency Identification (RFID) interrogator can read an Electronic Product Code (EPC) from an electronic tag.

This could be used to identify single consumer products, cases of products, pallets or even larger containers. In fact it is already in use within the distribution system. The advantage of using this within a retail store is that a trolley of goods could pass through a 'tunnel reader' and all of the tags read and products identified almost instantly, saving staff costs, speeding the checkout process and reducing theft. A further advantage would be that such a tag would not be limited to include only the GTIN of a bar code. The disadvantage is the cost of identifying every product in this way.

For the distribution of a pallet of different cases, the use of such a single tag to track where it is, to identify all its contents and the route it should be taking is very cost effective. To apply such technology to every single can and packet on that pallet is much more difficult to justify. Yet for many years it has been assumed that tags would be developed which were not attached to products but were actually included within the printing of the product label and that costs would come down to make this worthwhile.

The second technology is that of photographic imaging. Who would have believed twenty years ago what can now be achieved by the use of a mobile phone and the internet? So looking forward it must be possible to hold sufficient different images of a product such that a device can scan a particular can of baked beans and identify it from the digitised images it holds.

In other words a conveyor belt of products at a checkout could be identified from all these images rather than from the reading of a bar code.

So do I think the bar code will 'see me out'? I will give a politician's answer to that by saying it depends how long my shelf-life will be. I am sure a different technology will emerge as the replacement for the bar code. But the demise of the bar code has been talked about for many years. It has proved very resilient and, as we shall see below, is being adapted to accommodate new demands. But once some step change is introduced, it can transform the status quo very quickly.

The oldest man in the United Kingdom has died aged 110 whilst I have been writing this section of the book. He lived in Wirksworth, two hundred metres from my home. Unless I live to be a successor to him, I suspect that I will continue to see at least some bar codes in my local shops.

*****

If I do not see the number or code being replaced in the next few years, I do see some other significant changes. The pressures on retail costs and from customers must surely increase the use of self-scanning. It has been introduced and widely accepted at checkouts in a relatively short space of time. It seems logical to widen its use and to do so by reading bar codes whilst actually shopping. It is already happening in some stores. I am surprised it has not extended more rapidly. Of course there are security issues to be considered and a number of products such as alcohol and health related products may require staff approval to complete a purchase. But as customers have become comfortable using them at checkouts, I believe they will move to scanning during shopping in large numbers. Initially they will be provided with scanners by the store. However, since mobile phones already include bar code readers, is it possible that these phones could be used for in-store shop scanning too ?

Wherever the bar code is scanned, different bar codes will be seen. Once again this process has already begun. Although we have seen that the GTIN gives access to other information which does not need to be directly included in the code, some additional data may be required on each individual pack. For example, the need to ensure that a product is not sold beyond its 'sell by date' or from a batch with a manufacturing problem, would require a modified code. Therefore either longer codes or, more generally, codes stacked on top of each other will be seen. The requirement for this led to a group of symbols which were originally referred to as Reduced Space Symbology, reflecting the need to limit the printing space on a product. These extended and/or stacked symbols can already be seen and they are now named GS1 Databars.

Access to product information using <u>mobile phones</u> will be far more extensive. The ability to read a bar code in this way has already proved valuable to those with an allergy, say to nuts, who have been able to access the background product databases of manufacturers to ensure the product they are buying is safe for them. Since many other details about the ingredients, food additives and sources of products are already help within the Global Data Synchronisation Network, it seems logical to allow some form of access to customers.

The importance of shelf edge prices always matching the price held on a store computer (and charged to the customer) was emphasised in Part II. I had always assumed that the use of <u>electronic shelf edge prices</u> rather than printed labels would be more significant than it is.

This would have allowed prices to be controlled automatically using a computer system which had a plan of where products were located on each shelf in the store. It has not been used in many stores but now it could be achieved without electrical wiring by radio frequency transmission. Though I am less certain of its widespread use, I would expect to see much greater use in the future.

*****

Perhaps this list is too cautious. As you may have noted, all of the changes reflect ideas which are either already in some use or developing established technologies. Technological change must mean that further opportunities will emerge which I have not considered.

The number is a very simple basis for a system. The printed bar code with its numeric backup when a code cannot be read is also easily used and understood. So both adhere to the fundamental KISS principle of success – Keep It Simple Stupid. Overall I cannot see any fundamental discontent with the present system demanding change. Also I cannot see any excitement with an alternative system which would alter that perception.

As a result I believe that the significant changes which will occur will relate to practical operations rather than to the foundations of the current system.

However in conclusion I can make one prediction in which I have total confidence. My predictions will prove to be wrong.

25. Has It Been Worth It ?

Let me return to my parents' shops in order to give a personal answer to this question. I have already mentioned that their Fish and Chips shop of my teenage years has now become a Chinese takeaway. They bought it as an established business and sold it as one. As far as I have been able to establish it remained a Fish and Chips shop for many years after they left. However times and tastes change, and though the premises have continued to be used as some form of food takeaway, the products sold there are not the same.

I went back a few years ago to my parents' second shop which sold fruit and vegetables. It had been converted into a house. Once again my parents both bought and sold it as an established business. But its future looked bleak even at the time they sold it. The reason did not relate to any local change in the desire for different products. At the time of the purchase of the shop by my parents, they were unaware that the previous owner had already commenced to sell the same products to the same customers from a mobile van direct to their homes. This new form of competition resulted in an immediate reduction in their anticipated sales leading to a business which was no longer viable.

I have used these two examples to state the obvious. The retail offering is constantly changing both from new ownerships, new offerings and different forms of competition. In the six years I have lived in Wirksworth many shops have new owners ; a micro brewery, two fashion shops and a monthly farmers' market have started ; a long established builders' merchant and an estate agent have closed ; a carpet shop has changed to a paint shop, then to an antique shop, and now to a cafe. Within ten miles, a new supermarket has been built complete with a bypass to Matlock town centre, and two other supermarkets have changed ownership. Beyond all of that, online shopping has increased dramatically for a wide range of products as well as extending the reach of food supermarkets.

Within all of these changes the most significant overall development, at least in the United Kingdom, has been the growth and concentration in the power of a limited number of supermarket chains. The purpose of a chapter headed '*Has It Been Worth It?'* is not to discuss the wider merits and demerits of this changing retail environment. Much continues to be written on this subject and it arouses strong feelings. The supporters of this retail revolution point to the choice, constant availability, competitiveness and convenience of these stores. On the other hand the book 'Tescopoly' is, as the name suggests, about the Tesco supermarket chain and its position in British retailing which, whilst far from being a monopoly, is nevertheless more powerful than any such chain has ever achieved before. Needless to say, the book is critical of Tesco in terms of how it has achieved this position and its effect on other retailers, their own suppliers and the communities they serve. In Wirksworth there are no shops which are part of any store chain, other than the Spar shop I mentioned earlier. But if there were plans by any of the major supermarket chains to locate within the town, there would be polarised opinion with some welcoming and others campaigning to resist such a move.

However the retail revolution caused by article numbers and bar codes has also occurred within this period and cannot be divorced from the wider subject of the changing retail industry.

*****

You will have seen similar television programmes to the one I describe here. It was devoted to considering what effect the group of major supermarket chains had on a number of their suppliers. Among these was a farming family going back many generations, who were the latest of many to sell their herd of cows because the price demanded of them for their milk by supermarkets made their business no longer viable. And there was a lettuce producer who was being delisted by a supermarket chain after supplying them for over 20 years because they had found a cheaper alternative.

But what struck me most about this particular programme was its introduction. The image used as the lead-in to the programme was not supermarket logos, not their store fronts, not their interiors, not their customer trolleys piled high with goods, not empty High Street premises – no, it was ... bar codes. I believe that this association in the minds of the programme makers will be shared by at least some who have read this far. And it confirms my own thinking when planning this book that I should conclude it by assessing whether bar coding 'has been worth it'.

The whole basis of all that I have written so far has been that scanning has had a fundamental and widespread impact on retailing. I have emphasised that this is not simply a matter of increased efficiency at the checkout. The information provided has impacted almost every part of the operation, merchandise choice and space management, pricing and sales promotion, stock control and logistics management. It is true that some of these benefits have become available to independent retailers as the costs of scanning and personal computers have come down. There has been a trickle down effect. But there is little doubt that the major gains have been made by the largest retail chains and that these would not have been possible without the capture of the basic data provided by scanning.

That does not mean that bar codes are the sole reason for their success. Other developments in technology have been equally important. Computer processing power and demographic data have been used to determine where stores should be located and how large they should be. Storage, processing and transport methods have given opportunities to widen the ranges of fresh, chilled and frozen products. Improved transport and logistics networks have reduced the time taken to deliver products from a supplier to a store.

If that sounds like an apology for the bar code, it is not. Whilst a large part of my life has been intimately involved with this subject, I have spent as much time working in companies and organisations involved in the promotion of fair trade products, local sourcing and support for local retailers.

In both of these worlds, I have argued for efficiency and can see no virtue in limiting technical progress which allows this to happen. The use of scanning is clearly more efficient than the system of price stickers and price entry which it replaced. So the bar code has resulted in customers being served in all sizes of store more efficiently and at lower cost.

It has also provided raw data to give stores the opportunity to improve their supply processes, giving better service levels to customers and reducing waste.

Of course, the desire to improve the checkout process is not good news if you are the till operator who is made redundant. But it does not follow that unemployment increases as a result. The total economic picture is far more complex and I distrust analyses which simplify the issues by claiming fewer jobs or extra jobs as a result.

Of course the questions which are asked go beyond the changing number of checkout operators. Is the possibility of providing any product from anywhere at any time environmentally sustainable? Is the balance of power between retailers and their suppliers unhealthy? Is the consumer best served by the knowledge and power of large supermarket chains and the reduction in choice?

But I do not believe these questions negate in any way the benefits that have been made from the development and use of bar codes. Quite the opposite. The challenge now is to make even greater use of the data which comes from their use. And that will be the subject of my final section.

Has it been worth it? Of course it has. But can it become far more worthwhile? Of course it can.

*****

I will complete a quotation I referred to in Part III when talking about price promotions. The UK Member of Parliament John Denham has introduced a private member's bill. He has said, *'Supermarkets have a huge advantage over shoppers.*

While the supermarkets can base their pricing and promotion policies on sophisticated computer analysis of how we shop, most consumers are left shopping around and trying to work out value for money in much the same way they did decades ago. My bill will even things up a little, giving consumers more chance of really getting the best deal.'

The reason for his bill is because he believes many apparent bargains may not be what they seem. Therefore he would wish to force supermarkets to release pricing data, product by product and store by store, enabling shoppers with smartphones to scrutinise special offers, multibuys and 'buy one get one free' offers while pushing their trolleys around supermarkets.

I am not a typical shopper. This may be because of my own retail experience but is more likely to be the result of my mathematical mind and the fact that I have more time to shop. Finding a bargain for me does not involve assuming all special offers are good value because I always calculate the cost per kilogramme (if not provided), compare prices of different sizes and products, and look at the list of ingredients. Shopping with me cannot be much fun.

I had always assumed that a more educated, more knowledgeable population would shop increasingly as I do. My observations of other shoppers suggests that I am mistaken in that expectation. John Denham's bill will not be passed into law. But it does raise an important economic issue. Shopping has changed enormously. For many the days of bargaining about a price which I referred to in the first chapter of this book have gone long ago. Even the visit to a market where prices could be compared on different stalls is a thing of the past. They have been replaced by a single, buy everything in a single store, shopping visit and a belief in getting general value for money.

In this environment, Denham has echoed the theme of this book that bar coding has been part of this retail revolution. Retailers and their suppliers have used the data that has been produced to operate more efficiently and in more sophisticated ways. They will continue to produce more information from these data sources.

Customers have not kept pace with these changes. That does not imply that they have not also benefited from them. But there is a lot of data which is theoretically accessible to all. The combination of computer processing power and storage, the speed of the internet and the development of mobile technologies means that data on ingredients, sourcing, environmental impact, price comparison and history will be available. It is possible to envisage that this could be processed in such a way that it could be used by a shopper, if they wished to do so, at the point of deciding on a purchase. Far from being a threat to retailers, they may even be among the providers of the information.

Forty years ago the founders of the system emphasised that any potential success would need to ensure that there were benefits for retailers, manufacturers and customers. For customers they saw the big gain as being shorter supermarket queues and faster service as a result. There is now the opportunity to restate that position. Retailers will find new ways to make use of new technologies and processed data to improve their businesses both in store and behind the scenes. Manufacturers will do the same to source, produce and deliver in more efficient ways. And customers will see faster service and better information to aid their choice of products.

As a result I believe it is appropriate for me to conclude by giving you my special BOGOF (buy one get one free). You have paid for my answer to the question, *'Has it been worth it?'* which is *'Yes of course it has'.* Your free answer to the unasked question, *'Will it continue to be worth it?'* is *'Yes it will be too'.*

APPENDICES

- EAN/GS1 Membership
- GS1 Officers 1977 – 2013
- Where did EAN/GS1 meet
- Significant milestones of GS1 History
- EAN/GS1 first scanning stores
- Alphabetical list of EAN.UCC country prefix allocations

EAN/GS1 Membership

1977   Austria- Belgium & Luxembourg –Denmark – Finland – France – Germany – Italy - The Netherlands – Norway – Sweden – Switzerland - United Kingdom

1978   Spain - Japan

1979   Australia

1981   New Zealand

1982   South Africa – Serbia

1983   Czech Republic

1984   Hungary – Iceland - Israel

1985   Argentina – Brazil – Cyprus – Greece - Taiwan

1986   Portugal - Russian Federation

1987   Mexico – Singapore - Venezuela

1988   Malaysia - South Korea – Thailand - Turkey

1989   Chile – Colombia – Hong Kong -   Peru - Uruguay

1990   Cuba – Poland

1991   Bulgaria - Costa Rica, El Salvador, Guatemala, Honduras, Nicaragua and Panama - China

1992   Croatia – Ecuador – Ireland – Malta – Slovenia - Tunisia

1993   Estonia – Indonesia – Morocco - Philippines

1994 Algeria – Bolivia - Latvia- Lithuania – Macedonia – Mauritius – Paraguay – Romania – Slovakia – Ukraine

1995 Bosnia-Herzegovina - Dominican Republic – India – Moldova - Sri Lanka – Vietnam

1996 Armenia - Egypt - Georgia – Iran – Kazakhstan - Lebanon

1997 Belarus- Jordan

1998 Costa Rica - Guatemala- Nicaragua – Panama – Syria - Uzbekistan

1999 Azerbaijan - El Salvador- Honduras – Kenya – Macao - North Korea - Saudi Arabia

2000 Libya – Kuwait - United Arab Emirates

2001 Bahrain

2002 Kyrgyzstan – USA - Canada

2003 Cambodia - Mongolia

2006 Albania – Ghana - Ivory Coast

2007 Tajikistan – Nigeria - Pakistan

2008 Montenegro

2010 Senegal

2011 Brunei – Tanzania - Greece

GS1 Officers 1977 – 2013

Chairmen

1977 - 1988	Albert HEIJN	The Netherlands
1988 - 1991	Jean COLLIN	Belgium
1991 -1994	Roland FAHLIN	Sweden
1994 - 1997	Jacques A. N. VAN DIJK	The Netherlands
1997 - 2000	John BERRY	UK
2000 -2003	Laurie WILSON	Australia
2003 - 2006	Tim SMUCKER	USA
2006 - 2008	Danny WEGMAN	USA
2008 -2010	Robert A. McDONALD	USA
2011 - present	José LOPEZ	Spain

Secretary General EAN/President-CEO GS1

1977 - 1991	Etienne BOONET	Belgium
1992 - 2000	Reinhold VAN LENNEP	Belgium
2001 - 2003	Brian SMITH	New Zealand
2003 - present	Miguel LOPERA	Spain

Vice Chairmen

1977-1979	Jean VERMAELEN	Belgium
1979-1988	Jean COLLIN	Belgium
1988-1991	Roland FAHLIN	Sweden
1991-1994	Jacques A. N. VAN DIJK	The Netherlands
1994-1997	John BERRY	UK
1997-2000	Taha HUSSEINI	France
2000-2003	Peter JORDAN	UK
2003-2005	Peter JORDAN	UK
	& Juan-Antonio SANFELIU	Spain
2005-2007	Juan-Antonio SANFELIU	Spain
	& Seung-han LEE	South Korea
2007-2008	Juan-Antonio SANFELIU	Spain
	& Bob McDONALD	USA

2008-2009	Juan-Antonio SANFELIU	Spain
	& Zygmunt MIERDORF	Germany
2009-2010	Zygmunt MIERDORF	Germany
	& José LOPEZ	Spain
2010-2011	José LOPEZ	Spain
	& Tim SMUCKER	USA
2011-2012	Tim SMUCKER	USA
	& Zong-nan WANG	China

Where did EAN/GS1 meet

General Assembly

3/02/1977	Brussels, BELGIUM
2/09/1977	Brussels, BELGIUM
13/04/1978	Barcelona, SPAIN
5/04/1979	Vienna, AUSTRIA
2/03/1980	Roma, ITALY
22/05/1981	Stockholm, SWEDEN
7/05/1982	Chicago, USA
27/05/1983	Zürich, SWITZERLAND
2/12/1983	Paris, FRANCE
25/05/1984	Wiesbaden, GERMANY
10/05/1985	Auckland, NEW ZEALAND
30/05/1986	Barcelona, SPAIN
5/06/1987	Amsterdam, THE NETHERLANDS
20/05/1988	Tokyo, JAPAN
12/05/1989	Edinburgh, UK
18/05/1990	Copenhagen, DENMARK
19/04/1991	Melbourne, AUSTRALIA
11/10/1991	Stockholm, SWEDEN
21-22/05/1992	Brussels, BELGIUM
14/05/1993	Basel, SWITZERLAND
6/05/1994	Mexico City, MEXICO
26/05/1995	Reykjavik, ICELAND
10/05/1996	Lisbon, PORTUGAL
2/05/1997	Chicago, USA
15/05/1998	Sao Paulo, BRAZIL
11/05/1999	Berlin, GERMANY
24/05/2000	Beijing, CHINA
11/05/2001	Dublin, IRELAND
17/05/2002	Vienna, AUSTRIA
26/11/2002	Brussels, BELGIUM
28/05/2003	Amsterdam, THE NETHERLANDS
10/09/2003	Brussels, BELGIUM
12/05/2004	Chiangmai, THAILAND

17/02/2005	Brussels, BELGIUM
18/05/2005	Cape Town, SOUTH AFRICA
17/05/2006	St Julians, MALTA
23/05/2007	Seoul, SOUTH KOREA
21/05/2008	Dubrovnik, CROATIA
19/02/2009	Brussels, BELGIUM
13/05/2009	Santiago, CHILE
19/05/2010	Kuala Lumpur, MALAYSIA
18/05/2011	Paris, FRANCE
23/05/2012	Cartagena, COLOMBIA

Executive Committee

16/03/1977	Brussels, BELGIUM
9-10.05.1977	London, UK
11/08/1977	Brussels, BELGIUM
29-30.09.1977	Amsterdam, THE NETHERLANDS
15-16.12.1977	London, UK
20/03/1978	Paris, FRANCE
16/06/1978	Cologne, GERMANY
26/09/1978	Amsterdam, THE NETHERLANDS
1/12/1978	London, UK
16/02/1979	Brussels, BELGIUM
22/05/1979	Paris, FRANCE
20/09/1979	Munich, GERMANY
11/12/1979	London, UK
29/04/1980	Brussels, BELGIUM
29/08/1980	Copenhagen, DENMARK
31/10/1980	London, UK
10/03/1981	Amsterdam, THE NETHERLANDS
18/08/1981	Paris, FRANCE
27/11/1981	Vienna, AUSTRIA
17/03/1982	Vevey, SWITZERLAND
27/09/1982	Amsterdam, THE NETHERLANDS
07.12.1982	London, UK
24/03/1983	Cologne, GERMANY
9/09/1983	Salzburg, AUSTRIA
1/12/1983	London, UK
1/03/1984	Brussels, BELGIUM

13/09/1984	Copenhagen, DENMARK
7/12/1984	London, UK
20/03/1985	Montreux, SWITZERLAND
20/09/1985	Santa-Maria Ligure, ITALY
17/12/1985	Budapest, HUNGARY
24/03/1986	Jerusalem, ISRAEL
26/09/1986	Dubrovnic, YUGOSLAVIA
28/11/1986	Melbourne, AUSTRALIA
27/03/1987	Lisbon, PORTUGAL
11/09/1987	Helsinki, FINLAND
4/12/1987	Brussels, BELGIUM
18/03/1988	Florence, ITALY
28/10/1988	Oslo, NORWAY
13/03/1989	Buenos Aires, ARGENTINA
27/10/1989	Paris, FRANCE
9/03/1990	Cologne, GERMANY
19/10/1990	Vienna, AUSTRIA
15/02/1991	Budapest, HUNGARY
11/10/1991	Stockholm, SWEDEN

(The EC was disbanded in 1992 and replaced by a Board with a limited number of participants representing the industry and the MO's all over the community)

SIGNIFICANT MILESTONES of GS1 History

1973, 3 April

Industry leaders select a single standard for product identification (Universal Product Code symbol) over seven other options. Still in use today, this was the first GS1 bar code.

1974

Establishment of the Uniform Code Council (UCC) in the USA.

1974, 26 June

First live scan of a GS1 bar code on a Wrigleys gum in a Marsh supermarket in Ohio, USA.

1976

Based on the original GS1 bar code, a 13th digit was engineered allowing the Identification System to go global.

1977

Establishment of the European Article Numbering Association (GS1) as not for profit international association with Head Office in Brussels with 12 founding Member Organisations from European countries.
Launch of the common GS1 identification system to improve supply chain efficiency in the Retail sector.

1983

Expansion of GS1 standards beyond consumers units with ITF-14 bar codes for outer cases.

1989

Supply chain applications expansion of GS1 standards to logistics units with GS1-128 bar codes along with the GS1 Application Identifier published.

First step into eCom with the original version of the EANCOM Manual, an international standard for Electronic Data Interchange.

1990

The UCC and EAN International (GS1) sign a cooperative agreement formalising their intent to co-manage global standards. GS1 presence in 45 countries.

1995

GS1 expanding in the Healthcare sector with the first Healthcare Collaboration Project.

1996

Launch of SC31, the ISO committee for automatic identification and data capture standards.

1999

Launch of the Auto-ID Centre at the Massachusetts Institute of Technology, leading to the development of the Electronic Product Code GS1 DataBar (reduced space symbology) specifications approved.

2000

GS1 presence in 90 countries.

2002

Launch of Global Standards Management Process (GSMP)

2003

GS1 forms EPCglobal and initiates the development of the EPCglobal architecture and standards.
GS1 DataMatrix (first two dimensional symbol) approved.

2004

GS1 published the business message standards (using XML) and first standard for Radio-frequency identification (Gen2).
Launch of the Global Data Synchronisation Network (GDSN), a global, internet-based initiative that will enable trading partners to efficiently exchange product master data.

2005

Worldwide launch of GS1, the new name for the organisation.

2007

World Customs Organisation and GS1 sign a Memorandum of Understanding.
GS1 enters the world of Business-to-Consumer (B2C) solutions. The aim is to provide open standards to link product information with consumers and businesses through mobile devices.

2011

GS1 expands its' offering with the approval of the GS1 QR Code.

2013

GS1 celebrates 40 years as the Global Language of Business.
GS1 presence in 111 countries.

EAN/GS1 first Scanning stores

GS1 MO	Date	Store	City
Albania	January 2000	Big Market	Tirane
Algeria	1996	EDIPAL	Algiers
Argentina	August 1983	Supermarket Toledo	Mar del Plata
Armenia	1997	Supermarket Parma	Yerevan
Australia	1978	SIMS Supermarket	Melbourne/ Victoria
Austria	July 1978	MAXI MARKT	Linz
Azerbaijan	1998	Supermarket "Ramstore"	Baku
Bahrain	1989	JAWAD SUPERMARKETS	Nuwaidrat
Belarus	October 1993	Supermatket "Na Nemige"	Minsk
Belgium & Luxembourg	May 1981	Supermarkt Pollet	Westende (Belgium)
Bolivia	1994	Hipermaxi	Santa Cruz de la Sierra
Bosnia-Herzegovina	N.A.		
Brazil	1984	3 Stores: SENDAS S.A., BOMPRECO S.A. SUPERMERCADOS DO NORDESTE and WMS SUPERMERCADOS DO BRASIL LTDA	
Bulgaria	October 1992	METROCOM	Sofia
Canada	1975	Metro Stores	Montreal

Chile	1990	D&S - Supermercados Almac	Santiago
China	June 1992	Department Store in Liberation Road	Hangzhou (Zhejiang Province)
Colombia	1991	POMONA	Bogota
Costa Rica	October 1992	Más Por Menos (MXM) La Granja	San Pedro City
Croatia	Sept 1992	KRAŠ Retail Store	Zagreb
Cyprus	February 1981	Demos Supermarket	Nicosia
Czech Republic	1990	PRONTO PLUS	Prague
Denmark	1979	Strandberg Supermarked	Holbæk
Dominican Republic	1994	Ferretería Cuesta	Santo Domingo
Ecuador	April 1993	SUPERMAXI	Quito
El Salvador	1994	Almacenes Siman	San Salvador
Finland	1983	Elanto	Helsinki
France	June 1980	Casino	Saint Etienne
Germany	October 1977	Südmarkt	Augsburg
Greece	Sept 1992	Makro Cash & Carry Hellas (Metro Group)	Athens
Guatemala	1992	Supermercados Paiz	Guatemala
Honduras	1996	Supermercado la Colonia	Tegucigalpa
Hong Kong	March 1990	2 Stores: Park'n Shop and Wellcome	Hong Kong
Hungary	1984	SKÁLA METRO	Budapest
Iceland	April 1989	Bonus	

Country	Date	Store	City
India	2000	Foodworld' (a JV of Spencer's & Dairyfarm)	Chennai
Indonesia	May 1996	Hero Supermarket	Jakarta
Iran	1994	Sharvand Chain Stores	Tehran
Ireland	1989	L&N	Tramore, Co. Waterford
Italy	May 1983	Esselunga, Piazza Ovidio	Milano
Japan	1982	Seven Eleven Japan	Tokyo
Jordan	1990	CTown Store	Amman
Kenya	1996	UCHUMI SUPERMARKETS	Nairobi
Kyrgyzstan	July 2000	Beta Stores	Bishkek
Latvia	1993	Interpegro Latvia	Riga
Lithuania	1993	JSC VILBARA	Vilnius
Macedonia	1994	Supermarket TINEX	Skopje
Malaysia	October 1987	Parkson	Petaling Jaya
Malta	Dec 1993	Smart Supermarket	B'Kara
Mexico	Dec 1991	Aurrerá (Insurgentes)	Mexico City
Moldova	1995	FIDESCO	Chisinau
Mongolia	1998	NOMIN PLAZA	
Morocco	February 1990	Marjane	Rabat
New Zealand	1983	McDonalds Supermarket	Taradale, Hawkes Bay
Pakistan	Nov 2006	Makro	Lahore
Panama	8 July 1985	Supermaket GAGO	
Paraguay	October 1994	Supermercado Stock (for. Santa Isabel)	Asunción

264

Country	Date	Store	City
Philippines	Sept 1992	Uniwide Sales Warehouse Club	Quezon City
Poland	Sept 1993	Holam	Bielsko-Biala
Portugal	1985	2 stores: Pão de Açucar, Continente	Lisbon, Matosinhos
Romania	October 1996	Metro Cash&Carry Romania	Bucharest
Russian Federation	Sept 1987	Bolshevichka Apparel Store	Moscow
Singapore	Dec 1991	NTUC Fairprice Co-operative Ltd	Singapore
Slovakia	Sept 1991	20 petrol stations Benzinol s.p.	Zilina and surroudings
Slovenia	1985	Poslovni sistem Mercator	Ljubljana
South Africa	1982	SPAR	Pinetown
South Korea	May 1984	Shinsegae Department Store	Seoul
Spain	October 1982	MERCADONA	Valencia
Sweden	April 1982	ICA Supermarket Västerhallen	Enköping
Switzerland	1987	Migros Supermarket	San Antonino (Canton of Tessin)
Taiwan	1989	7-11 Chain Stores	Taipei
Thailand	1988	Central Department Store	Bangkok
The Netherlands	January 1977	Albert Heijn	Heemskerk
Tunisia	March 1994	MONOPRIX	Tunis
United Kingdom	October 1979	Keymarkets	Spalding
Uruguay	Sept 1993	Tienda Inglesa	Montevideo
USA	June 1974	Marsh's Supermarket	Troy (Ohio)

Uzbekistan	1995	MIR store	Tashkent
		Supermercados	
Venezuela	1988	Victoria	Caracas
	October	Minimart	Ho Chi Minh
Vietnam	1993	Supermarket	city
Yugoslavia/Se			
rbia	1985	Centroprom	Belgrade

Alphabetical list of EAN.UCC country prefix allocations

Assignment of prefix digits, alphabetically			
Country or Use	Prefix	Country or Use	Prefix
Algeria (GS1 Algeria)	613	Latvia (GS1 Latvia)	475
Argentina (GS1 Argentina)	779	Lebanon (GS1 Lebanon)	528
Armenia (GS1 Armenia)	485	Libya (GS1 Libya)	624
Australia (GS1 Australia)	93	Lithuania (GS1 Lithuania)	477
Azerbaijan (GS1 Azerbaijan)	476	Luxembourg & Belgium (GS1	54
Austria (GS1 Austria)	90-91	Luxembourg & Belgium)	
Bahrain (GS1 Bahrain)	608	Macau (GS1 Macau)	958
Belarus (GS1 Belarus)	481	Macedonia (GS1 Macedonia)	531
Belgium & Luxembourg (GS1	54	Malaysia (GS1 Malaysia)	955
Luxembourg & Belgium)		Malta (GS1 Malta)	535
Bolivia (GS1 Bolivia)	777	Mauritius (GS1 Mauritius)	609
Bosnia-Herzegovina (GS1 BIH)	387	Mexico (GS1 Mexico)	750
Books (ISBN)	978-979	Moldova (GS1 Moldova)	484
Brazil (GS1 Brasil)	789	Mongolia (GS1 Mongolia)	865
Bulgaria (GS1 Bulgaria)	380	Morocco (GS1 Maroc)	611
Cambodia (GS1 Cambodia)	884	Netherlands (GS1 Nederlands)	87
Canada & USA (GS1 US)	00-13	New Zealand (GS1 New Zealand)	94
Chile (GS1 Chile)	780	Nicaragua (GS1 Nicaragua)	743
China (GS1 China)	690-693	North Korea (GS1 North Korea)	867
Colombia (GS1 Colombia)	770	Norway (GS1 Norge)	70
Common currency coupons	981-982	Panama (GS1 Panama)	745
Costa Rica (GS1 Costa Rica)	744	Paraguay (GS1 Paraguay)	784
Coupons	98-99	Periodicals (ISSN)	977
Croatia (GS1 Croatia)	385	Peru (GS1 Peru)	775
Cuba (GS1 Cuba)	850	Philippines (GS1 Philippines)	480
Cyprus (GS1 Cyprus)	529	Poland (GS1 Poland)	590
Czech (GS1 Czech Republic)	859	Portugal (GS1 Portugal)	560
Denmark (GS1 Denmark)	57	Refund receipts	980
Ecuador (GS1 Ecuador)	786	Republica Dominicana (GS1 RD)	746
Egypt (GS1 Egypt)	622	Restricted Circulation Numbers	20-29
El Salvador (GS1 El Salvador)	741	Romania (GS1 Romania)	594
Emirates (GS1 Emirates)	629	Russian Federation (GS1 Russia)	46
Estonia (GS1 Estonia)	474	Saudi Arabia (GS1 Saudi Arabia)	628
Finland (GS1 Finland)	64	Singapore (GS1 Singapore)	888
France (GS1 France)	30-37	Slovakia (GS1 Slovakia)	858
Georgia (GS1 Georgia)	486	Slovenia (GS1 Slovenia)	383
Germany (CCG)	400-440	South Africa (GS1 South Africa)	600-601
Greece (GS1 Greece)	520	South Korea (GS1 Korea)	880
Guatemala (GS1 Guatemala)	740	Spain (GS1 Spain)	84
Honduras (GS1 Honduras)	742	Sri Lanka (GS1 Sri Lanka)	479
Hong Kong (GS1 Hong Kong)	489	Sweden (GS1 Sweden)	73

Hungary (GS1 Hungary)	599	Switzerland (GS1 Switzerland)	76
Iceland (GS1 Iceland)	569	Syria (GS1 Syria)	621
India (GS1 India)	890	Taiwan (GS1 Taiwan)	471
Indonesia (GS1 Indonesia)	899	Thailand (GS1 Thailand)	885
Iran (GS1 Iran)	626	Tunisia (GS1 Tunisia)	619
Ireland (GS1 Ireland)	539	Turkey (GS1 Turkey)	869
Israel (GS1 Israel)	729	Ukraine (GS1 Ukraine)	482
Italy (GS1 Italy)	80-83	United Kingdom (GS1 UK)	50
Japan (GS1 Japan)	45 & 49	Uruguay (GS1 Uruguay)	773
Jordan (GS1 Jordan)	625	USA & Canada (GS1 US)	00-13
Kazakhstan (GS1 Kazakhstan)	487	Uzbekistan (GS1 Uzbekistan)	478
Kenya (GS1 Kenya)	616	Venezuela (GS1 Venezuela)	759
Kyrgyzstan (GS1 Kyrgyzstan)	470	Vietnam (GS1 Vietnam)	893
Kuwait (GS1 Kuwait)	627	Yugoslavia (GS1 Yugoslavia)	860

Numerical listing of EAN•UCC prefix allocations

Assignment of prefix digits, numerically			
Country or Use	Prefix	Country or Use	Prefix
Canada & USA (GS1 US)	00-13	Kuwait (GS1 Kuwait)	627
Restricted Circulation Numbers	20-29	Saudi Arabia (GS1 Saudi Arabia)	628
France (GS1 France)	30-37	Emirates (GS1 Emirates)	629
Bulgaria (GS1 Bulgaria)	380	Finland (GS1 Finland)	64
Slovenia (GS1 Slovenia)	383	China (GS1 China)	690-693
Croatia (GS1 Croatia)	385	Norway (GS1 Norway)	70
Bosnia-Herzegovina (GS1 Bosnia-Herzegovina)	387	Israel (GS1 Israel)	729
		Sweden (GS1 Sweden)	73
Germany (GS1 Germany)	400-440	Guatemala (GS1 Guatemala)	740
Japan (GS1 Japan)	45 & 49	El Salvador (GS1 El Salvador)	741
Russian Federation (GS1 Russia)	460-469	Honduras (GS1 Honduras)	742
		Nicaragua (GS1 Nicaragua)	743
Kyrgyzstan (GS1 Kyrgyzstan)	470	Costa Rica (GS1 Costa Rica)	744
Taiwan (GS1 Taiwan)	471	Panama (GS1 Panama)	745
Estonia (GS1 Estonia)	474	Republica Dominicana (GS1 RD)	746
Latvia (GS1 Latvia)	475	Mexico (GS1 Mexico)	750
Azerbaijan (GS1 Azerbaijan)	476	Venezuela (GS1 Venezuela)	759
Lithuania (GS1 Lithuania)	477	Switzerland (GS1 Switzerland)	76
Uzbekistan (GS1 Uzbekistan)	478	Colombia (GS1 Colombia)	770
Sri Lanka (GS1 Sri Lanka)	479	Uruguay (GS1 Uruguay)	773
Philippines (GS1 Philippines)	480	Peru (GS1 Peru)	775
Belarus (GS1 Belarus)	481	Bolivia (GS1 Bolivia)	777
Ukraine (GS1 Ukraine)	482	Argentina (GS1 Argentina)	779
Moldova (GS1 Moldova)	484	Chile (GS1 Chile)	780
Armenia (GS1 Armenia)	485	Paraguay (GS1 Paraguay)	784
Georgia (GS1 Georgia)	486	Ecuador (GS1 Ecuador)	786
Kazakhstan (GS1 Kazakhstan)	487	Brazil (GS1 Brasil)	789
Hong Kong (GS1 Hong Kong)	489	Italy (GS1 Italy)	80-83
Japan (GS1 Japan)	49 & 45	Spain (GS1 Spain)	84
United Kingdom (GS1 UK)	50	Cuba (GS1 Cuba)	850
Greece (GS1 Greece)	520	Slovakia (GS1 Slovakia)	858
Lebanon (GS1 Lebanon)	528	Czech (GS1 Czech Republic)	859
Cyprus (GS1 Cyprus)	529	Yugoslavia (GS1 Yugoslavia)	860
Macedonia (GS1 Macedonia)	531	North Korea (GS1 North Korea)	865
Malta (GS1 Malta)	535	Mongolia (GS1 Mongolia)	867
Ireland (GS1 Ireland)	539	Turkey (GS1 Turkey)	869
Belgium & Luxembourg (GS1 Luxembourg & Belgium)	54	Netherlands (GS1 Nederlands)	87
		South Korea (GS1 Korea)	880
Portugal (GS1 Portugal)	560		

Iceland (GS1 Iceland)	569	Cambodia (GS1 Cambodia)	884
Denmark (GS1 Danmark)	57	Thailand (GS1 Thailand)	885
Poland (GS1 Poland)	590	Singapore (GS1 Singapore)	888
Romania (GS1 Romania)	594	India (GS1 India)	890
Hungary (GS1 Hungary)	599	Vietnam (GS1 Vietnam)	893
South Africa (GS1 South Africa)	600-601	Indonesia (GS1 Indonesia)	899
		Austria (GS1 Austria)	90-91
Bahrain (GS1 Bahrain)	608	Australia (GS1 Australia)	93
Mauritius (GS1 Mauritius)	609	New Zealand (GS1 New Zealand)	94
Morocco (GS1 Morocco)	611	Malaysia (GS1 Malaysia)	955
Algeria (GS1 Algeria)	613	Macau (GS1 Macau)	958
Kenya (GS1 Kenya)	616	Periodicals (ISSN)	977
Tunisia (GS1 Tunisia)	619	Books (ISBN)	978-979
Syria (GS1 Syria)	621	Refund receipts	980
Egypt (GS1 Egypt)	622	Common currency coupons	981-982
Libya (GS1 Libya)	624	Coupons	99
Jordan (GS1 Jordan)	625		
Iran (GS1 Iran)	626